Lecture Notes in Mathematics

Edited by A. Dold and B. Eckmann

1150

Bernhard Kawohl

Rearrangements and Convexity of Level Sets in PDE

Springer-Verlag
Berlin Heidelberg New York Tokyo

Author

Bernhard Kawohl
Universität Heidelberg, Sonderforschungsbereich 123
Im Neuenheimer Feld 294, 6900 Heidelberg, Federal Republic of Germany

Mathematics Subject Classification (1980): 26B25, 26D10, 35A15, 35B05, 35B50, 35J20, 35J25, 35J65, 49G05

ISBN 3-540-15693-3 Springer-Verlag Berlin Heidelberg New York Tokyo
ISBN 0-387-15693-3 Springer-Verlag New York Heidelberg Berlin Tokyo

Library of Congress Cataloging in Publication Data. Kawohl, Bernhard, 1952- Rearrangements and some maximum principles in PDE. (Lecture notes in mathematics; 1150) Bibliography: p. Includes index. 1. Differential equations, Partial. 2. Maximum principles (Mathematics) I. Title. II. Series: Lecture notes in mathematics (Springer-Verlag); 1150. QA3.L28 no. 1150 [QA374] 510 s [515.3'53] 85-20806 ISBN 0-387-15693-3

Printing and binding: Beltz Offsetdruck, Hemsbach/Bergstr.
2146/3140-543210

dedicated to my family

CONTENTS

I. INTRODUCTION

These notes have their origin in a conjecture of J. Rauch. Let Ω be a domain in $\mathbb{R}^n$ and let ν_2 be the smallest positive eigenvalue of the Laplace operator in Ω under homogeneous Neumann boundary conditions. Then, so the conjecture, the associated eigenfunction u_2 should attain its maximum and minimum on the boundary $\partial\Omega$. More on the conjecture can be found in § 5.

The conjecture and attempts to prove it led the author of these notes to questions of the type: Suppose u solves the semilinear differential equation $\Delta u + f(u) = 0$ in a domain Ω . What is known about the critical points of u , i.e. the points in which ∇u , the gradient of u vanishes? What is known about the shape of u ? How does the shape of Ω influence the shape of u , if u satisfies boundary conditions on $\partial\Omega$? If Ω is convex, does u have convexity properties? If Ω is symmetric, does u have symmetry properties?

The answer to these questions is not easy, as can be seen from eigenfunctions of the Laplace operator in a circular plane domain.

There are various methods to approach problems of this type, and two if them are presented here.

The first method is the rearrangement method in which one replaces a given function u by a related function u^* which has some desired properties like symmetry or monotonicity. The function u^* can be reconstructed from its level sets $\Omega_c^* := \{x \in \Omega \mid u^*(x) \geq c\}$, just like a threedimensional mountain can be reconstructed from a map that shows all of its level lines or lines of constant height. Let us stick to this comparison and compare u and u^* with a hilly region before and after landscaping. In landscaping we have choices, in rearrangement methods, too. Once we decide which rearrangement to choose, one of the standard application of rearrangement arguments occurs in the calculus of variations along the following lines: Suppose that a function u minimizes a variational functional $J(v)$ over a set $\mathbb{K}$ of admissible functions. Then replace u by u^* , show that $u^* \in \mathbb{K}$ and that $J(u^*) < J(u)$ unless $u = u^*$. If u is the unique minimizer it suffices to derive $J(u^*) \leq J(u)$ to conclude $u = u^*$.

The second method consists of various maximum principles. They can and will be used to prove convexity- and symmetry-properties of solutions to differential equations.

Both methods are global in nature, and they have to be global because the shape of solutions to differential equations is a global property.

Both methods are fairly elementary in the sense that any graduate student with a little knowledge in variational inequalities and calculus should be able to understand the proofs. Aside from Federer's coarea formula there will be no use of geometric measure theory.

Both methods work in arbitrary dimensions!

Both methods have been used before, but we have been able to extend them to considerably wider classes of problems. Wellknown texts on rearrangement are the books by G. Polya and G. Szegö [156] and C. Bandle [19] and the papers of G. Talenti [183, 184]. Powerful applications of the maximum principle are due to B. Gidas, W.M. Ni and L. Nirenberg [85] and N. Korevaar [112].

Let us briefly explain our contribution. There are many papers on various rearrangements scattered in the mathematical literature and in these notes we have attempted to reference most and survey some of them, to show their common aspects and to give a unified exposition of their main properties. It is unavoidable that some of our results represents mathematical folklore, e.g. the observation that Steiner symmetrization works not only on rectangles in $\mathbb{R}^2$ but also on cylindrical domains in $\mathbb{R}^n$. Nevertheless some of the allegedly evident statements referring to rearrangements have never been properly proved. To give an example, the passage from Steiner to Schwarz symmetrization in [180] is a delicate procedure and not as evident as generally believed, see the end of § 8.

To familiarize the reader with technical details we included some paragraphs (3 and 4) on rearrangement of a function of a single variable. In one dimension these technical complications occur on a smaller scale and we hope that this exposition simplifies the understanding of the higher dimensional case.

In the authors opinion there are three main results on rearrangements in these notes. One is the discussion of the equality sign in the inequality

$$\int_{\Omega} |\nabla u|^2 \, dx \geq \int_{\Omega^*} |\nabla u^*|^2 \, dx \quad ,$$

and in similar inequalities. While G. Polya and G. Szegö wrote that there is "little hope for discussing the sign of equality" [156, p. 186], E. Lieb suggested that "probably a strict version ... is true" [116, p. 97], meaning that equality would imply $u = u^*$ modulo translation. We shall show that both guesses were appropriate, since there are large classes of functions for which one can or cannot discuss the equality sign†). This result is applied to different rearrangements and to different problems and has some nice consequences. It provides a proof of J. Rauch's conjecture (§ 5), and of the sharpness of the Krahn-Faber inequality $\lambda_1(\Omega) \geq \lambda_1(\Omega^*)$ (§ 8). It can also be used to increase the known multiplicity of solutions to a semilinear equation on an annulus and to prove an oscillation theorem for variational solutions to the plasma problem on an annulus (§ 9).

The second main result is the rediscovery of monotone decreasing rearrangement (§ 5) and its applicability to free boundary problems. In particular one can use it to derive the statement that a free boundary has to be a graph of a function of a certain $(n-1)$ dimensional variable.

The third main result is the observation that even though under star-shaped rearrangement one does not have Cavalieri's principle

$$\int_{\Omega} F(u) \, dx = \int_{\Omega^*} F(u^*) \, dx \quad ,$$

there is a substitute for it, an inequality (§ 6). This can be used to derive new Lipschitz-regularity results for some free boundary problems. Our results on Schwarz- and circular symmetrization are derived from corresponding ones for Steiner symmetrization (§ 7).

In the course of these investigations we also prove the orderedness of minimizing solutions of a variational problem, which seems to be known as jet and cavity problem.

Our contributions to maximum principles are less voluminous. First we derive two consequences of the main result of B. Gidas, W.M. Ni and L. Nirenberg (§ 10).

---

†) *After this manuscript had been sent to the publisher for evaluation the author learned of the related paper [76].*

Then we extend a method of Gabriel and Lewis to prove convexity of level sets $\Omega_c := \{x \in \Omega | \; u(x) \geq c\}$ of solutions u to some exterior free boundary problems. A difficulty in these problems is that the solutions are not classical, i.e. not of class C^2, so that classical maximum principles cannot be applied right away (§ 11). In addition the second part of these notes contains a list of tricks on how to prove convexity of level sets.

Finally we give an extension of N. Korevaar's concavity maximum principle [112] in § 12. This principle consists of two parts. In his boundary point lemma we can drop three out of six assumptions and weaken a fourth. In his maximum principle theorem we could considerably weaken a concavity assumption on a quasilinear term b [99]. Independently this was done by A. Kennington, too.

As a consequence of our concavity maximum principle we can answer an open problem of L. Caffarelli and A. Friedman on the asymptotic location of the plasma in a problem from plasma physics.

Of course there are other methods to prove statements about the "shape" of solutions to differential equations. For semilinear parabolic equations we mention results of K. Nickel [145], H. Matano [134], W.M. Ni and P. Sacks [143] and W. Walter [198] and admit that we have ignored group theoretic methods.

There are interesting relations between the geometry of a domain Ω and the existence or stability of solutions to semilinear elliptic and parabolic boundary problems. If we consider the simple problem

$$\begin{aligned} \Delta u + u^p &= 0, u > 0 \quad &&\text{in } \Omega \\ u &= 0 &&\text{on } \partial\Omega \end{aligned}$$

in a bounded domain $\Omega \subset \mathbb{R}^n$ and for $p > 1$, then it is known

a) that there exists no solution if $n > 2$, Ω is starshaped with respect to a point and $p \geq \frac{n+2}{n-2}$ [153],

b) there exists a unique solution if Ω is a ball in $\mathbb{R}^n$ and $p < \frac{n+2}{n-2}$ [85],

c) there is e.g. more than one solution if Ω is on annulus in $\mathbb{R}^n$ and $p < \frac{n+2}{n-2}$ [38b].

Similar relations are still being discovered, we refer to [59, 67, 119, 170]. Of particular interest are convex domains. One can show that on those domains the only stable solutions to

$$\Delta u + f(u) = 0 \quad \text{in} \quad \Omega$$

$$\frac{\partial u}{\partial n} = 0 \quad \text{on} \quad \partial\Omega$$

have to be constant [47-49, 108, 135].

The methods that are described in these notes will be applied to various problems from partial differential equations and the calculus of variations. We shall not treat questions of existence and regularity of solutions to these problems, but instead give references to the existing literature from which they were taken. We assume that the reader is familiar with or willing to look up the notions of Sobolev spaces, weak solutions to partial differential equations and maximal monotone operators. These can be found in the standard literature, e.g. [3, 39, 142].

Finally we want to explain some notation. Throughout these notes $\Omega \subset \mathbb{R}^n$ and $\Omega' \subset \mathbb{R}^{n-1}$ denote bounded domains; u, v, w etc. are real valued functions on $\bar{\Omega}$. $\mathbb{R}^+ := (0,\infty)$, $\mathbb{R}^+_0 := [0,\infty)$, $m_n(D)$ denotes n-dimensional Lebesgue measure of a Lebesgue measurable set $D \subset \mathbb{R}^n$. All integrals are understood as Lebesgue integrals, unless otherwise indicated. $\chi_A(x)$ denotes the characteristic function of a set $A \subset \mathbb{R}^n$; it is 1 for $x \in A$ and 0 for $x \notin A$, $d(x,A)$ denotes the distance of x to A, so $d(x,A) = \inf\{|x-y| \mid y \in A\}$, where $|\ |$ stands for the Euclidean distance. Correspondingly $d(A,B) = \inf\{|y-x| \mid x \in A, y \in B\}$ for subsets A and B of $\mathbb{R}^n$. The support of a nonnegative function $u : \bar{\Omega} \to \mathbb{R}$ is denoted by $\text{supp}\ u = \overline{\{x \in \bar{\Omega} \mid u(x) > 0\}}$, the gradient of u by ∇u, the Laplace operator by Δ. $W^{k,p}(\Omega)$ and $C^{k,\alpha}(D)$, $k \in \mathbb{R}$, $1 \leq p \leq \infty$, $0 < \alpha \leq 1$ are the commonly used notations for Sobolev spaces, i.e. spaces of functions whose derivatives up to the order k are in $L^p(\Omega)$, and for spaces of functions whose derivatives up to order k are Hölder continuous. The space $C^{k,0}(D)$ is also denoted by $C^k(D)$. The boundary $\partial\Omega$ of a domain $\Omega \in \mathbb{R}^n$ is called of class $C^{k,\alpha}$ if it has a local representation as a $C^{k,\alpha}$ function of n-1 variables. The symbol * will refer to different rearrangements and there will be different definitions of smooth functions in different paragraphs, each of them being valid within a paragraph. This was done on purpose to demonstrate that the same basic ideas can be applied to different rearrangements.

In various places we shall mention piecewise linear functions in a bounded domain $\Omega \subset \mathbb{R}^n$. We call a function $u : \overline{\Omega} \to \mathbb{R}$ piecewise linear if Ω can be divided into finitely many sets D_j such that u is affine in the interior of each such set D_j. Furthermore the boundary ∂D_j of each set D_j is contained in the union of $\partial\Omega$ and finitely many $(n-1)$ dimensional hyperplanes. If $d(D_j, \partial\Omega)$ is positive this means thad D_j is polyhedral. For the readers convenience we have added a small index of examples and assumptions which are frequently referred to. This can be found on the last page.

Most of the material in these notes was given in a special topics course at Brown University in 1984. I am gratefully indebted to various friends, colleagues and superiors; in particular to C. BANDLE for here supportive interest in this work, to H. GRABMÜLLER for many discussions and helpful criticism, to the Institute für Angewandte Mathematik of the Universität Erlangen-Nürnberg and to the Lefschetz Center of Brown University for their respective working conditions, as well to C. DAFERMOS, J.I. DIAZ, L.C. EVANS, E. GIARUSSO, J. HALE, M. MARCUS, G. KEADY, A. KENNINGTON, L.E. PAYNE, G. TALENTI and J.L. VAZQUEZ for stimulating discussions and correspondence.

This work was financially supported in part by the Deutsche Forschungsgemeinschaft (DFG).

II. REARRANGEMENTS

II.1 A catalogue of rearrangements

Throughout these notes we shall in general only rearrange Lipschitz continuous functions $u : \mathbb{R}^n \to \mathbb{R}_o^+ = [0,\infty)$ with bounded support $D \subset \mathbb{R}^n$ or Lipschitz continuous functions $u : \overline{\Omega} \to \mathbb{R}$ which are defined on compact sets $\overline{\Omega} \subset \mathbb{R}^n$.

The common feature of all rearrangements is that a given function u is transformed into a new function u^* , and that u^* has some desired property, e.g. symmetries. This is done by a rearrangement of the level sets $\Omega_c := \{x \in \overline{\Omega} | \ u(x) \geq c\}$, $c \in \mathbb{R}$, of u . Therefore the rearrangement of a function u is closely tied to the rearrangement of subsets of $\mathbb{R}^n$.

If $D \subset \mathbb{R}^n$ is a Lebesgue measurable set we denote its rearrangement by D^* . The symbol $*$ will be used for different kinds of rearrangements. There will be no confusion about its meaning as long as it is clear from the context. We shall also introduce the notation $D^{*(p)}$ with $p \in \mathbb{R}$ for some distinguished rearrangements.

If $D = \emptyset$, then $D^* = \emptyset$ by definition for all kinds of rearrangements.

One we define the rearrangement D^* of a Lebesgue measurable set D , we can also define the rearrangement u^* of a function u as follows.

$$u^*(x) \quad := \quad \sup\{c \in \mathbb{R} | \ x \in \Omega_c^*\} \quad \text{for} \quad x \in \overline{\Omega} \quad := \quad (\overline{\Omega})^* \quad . \qquad (2.1)$$

Notice that this definition is meaningful for Lebesgue measurable functions on $\mathbb{R}^n$. In fact many of the results of this chapter can be stated for this large class of functions. The restriction to Lipschitz continuous functions on compact domains has however several reasons.

i) We want to apply our results to solutions of variational problems in bounded domains. Those solutions are known to be Lipschitz continuous and therefore we do not need to formulate our theorems in their utmost generality. This would only obscure some essential ideas.

ii) We intend to look at functionals of the type $\int_\Omega G(|\nabla u|)\,dx$ and investigate their behaviour under rearrangement of u. Here $G: \mathbb{R}_o^+ \to \mathbb{R}$ is convex. Our assumption on the Lipschitz continuity of u ensures that the functional is finite. Otherwise we would always have to add extra statements on the case that these integrals are infinite.

iii) There are several slightly different definitions of u^* in the literature, e.g. $u^*(x) = \inf \{c \in \mathbb{R}_o^+ \,|\, x \notin \bar{\Omega}_c^*\}$ for $x \in \bar{\Omega}^*$, for nonnegative functions with compact support $\bar{\Omega}$ in $\mathbb{R}^n$, or definitions using the open level sets $\{x \in \Omega \,|\, u(y) > c\}$, $c \in \mathbb{R}$, $\Omega \subset \mathbb{R}^n$ open. For continuous functions u^* these definitions are equivalent to ours, but for measurable functions they are not.

Let $D \subset \mathbb{R}^n$ be a compact (and hence Lebesgue measurable) set and let $m_n(D)$ denote the Lebesgue measure of D. For $n = 1$ we present three ways to rearrange D into a interval of length $m_1(D)$.

a) The <u>monotone rearrangement</u> D^* of D, which is defined by

$$D^* = \begin{cases} \{x \in \mathbb{R}_o^+ \,|\, x \leq m_1(D)\} & \text{if } D \neq \emptyset, \\ \emptyset & \text{if } D = \emptyset. \end{cases}$$

b) The <u>symmetric rearrangement</u> D^* of D, given by

$$D^* = \begin{cases} \{x \in \mathbb{R} \,|\, |x| \leq \frac{1}{2} m_1(D)\} & \text{if } D \neq \emptyset, \\ \emptyset & \text{if } D = \emptyset. \end{cases}$$

c) The <u>quasiconcave rearrangement</u> D^* of D, defined by

$$D^* = \begin{cases} \{x \in \mathbb{R} \,|\, -m_1(D \cap \mathbb{R}^-) \leq x \leq m_1(D \cap \mathbb{R}^+)\} & \text{if } D \neq \emptyset, \\ \emptyset & \text{if } D = \emptyset. \end{cases}$$

If $u : \mathbb{R} \to \mathbb{R}$ is a Lipschitz continuous function with compact support $\overline{\Omega}$, or if $\overline{\Omega} \subset \mathbb{R}$ is a compact interval and $u : \overline{\Omega} \to \mathbb{R}$ is Lipschitz continuous we can formally define

$$u^*(x) := \sup \{c \in \mathbb{R} \mid x \in \Omega_c^*\} \quad \text{for} \quad x \in \overline{\Omega}^* \tag{2.1}$$

and call the function $u^*(x)$:

a) the monotone decreasing rearrangement of u if Ω_c^* and $\overline{\Omega}^*$ are the monotone rearrangements of Ω_c and $\overline{\Omega}$,

b) the symmetric decreasing rearrangement of u if Ω_c^* and $\overline{\Omega}^*$ are the symmetric rearrangements of Ω_c and $\overline{\Omega}$, and

c) the quasiconcave rearrangement of u if Ω_c^* and $\overline{\Omega}^*$ are the quasiconcave rearrangements of Ω_c and $\overline{\Omega}$. In the sequel we shall only refer to quasiconcave rearrangement under the additional assumption that $0 \in \overline{\Omega}$ and that u attains its maximum in the origin. The reason for this restriction can be illustrated in form of the following example, in which u does not attain its maximum at zero, and where u^* is discontinuous.

Example 2.1: Let $\overline{\Omega} := [-1,2]$ and let

$$u(x) := \begin{cases} x, & \text{if } -1 \le x \le 1, \\ 2-x, & \text{if } 1 \le x \le 2. \end{cases}$$

Then the three rearrangements of u are given by:

a)
$$u^*(x) = \begin{cases} 1 - \frac{1}{2}x & \text{if } 0 \le x \le 2, \\ 2 - x & \text{if } 2 \le x \le 3, \end{cases} \qquad x \in [0,3] \;;$$

b)
$$u^*(x) = \begin{cases} 1 - |x| & \text{if } |x| \le 1, \\ 2 - 2\;x & \text{if } 1 \le |x| \le 1.5, \end{cases} \qquad x \in [-1.5,1.5] \;;$$

c)
$$u^*(x) = \begin{cases} x & \text{if } -1 \le x \le 0, \\ 1 - \frac{1}{2}x & \text{if } 0 \le x \le 2, \end{cases} \qquad x \in [-1,2] \,.$$

For the readers convenience these functions are plotted in Figure 2.1 below.

Remark 2.1 On the relation between a) and c).

Under the assumption that $u(o) = \max_{x \in \bar{\Omega}} u(x)$ one can relate the decreasing rearrangement a) to quasiconcave rearrangement c). Let us restrict u to $\bar{\Omega} \cap \mathbb{R}_o^+$, denote this restriction by $u_1(x)$ and form the monotone decreasing rearrangement $u_1^*(x)$. The function $u_1^*(x)$ is the restriction of the quasiconcave rearrangement u^* of u to $\bar{\Omega}^* \cap \mathbb{R}_o^+$.

Remark 2.2 On the relation between b) and c).

Under the assumption $u(x) = u(-x)$ the quasiconcave rearrangement c) and the symmetric decreasing rearrangement b) coincide. To generate the rearrangement u^* it suffices to proceed as in Remark 2.1 and form the monotone decreasing rearrangement u_1 of the restriction of u to nonnegative $x \in \bar{\Omega}$. Then u^* is the symmetric extension of u_1^* across the origin.

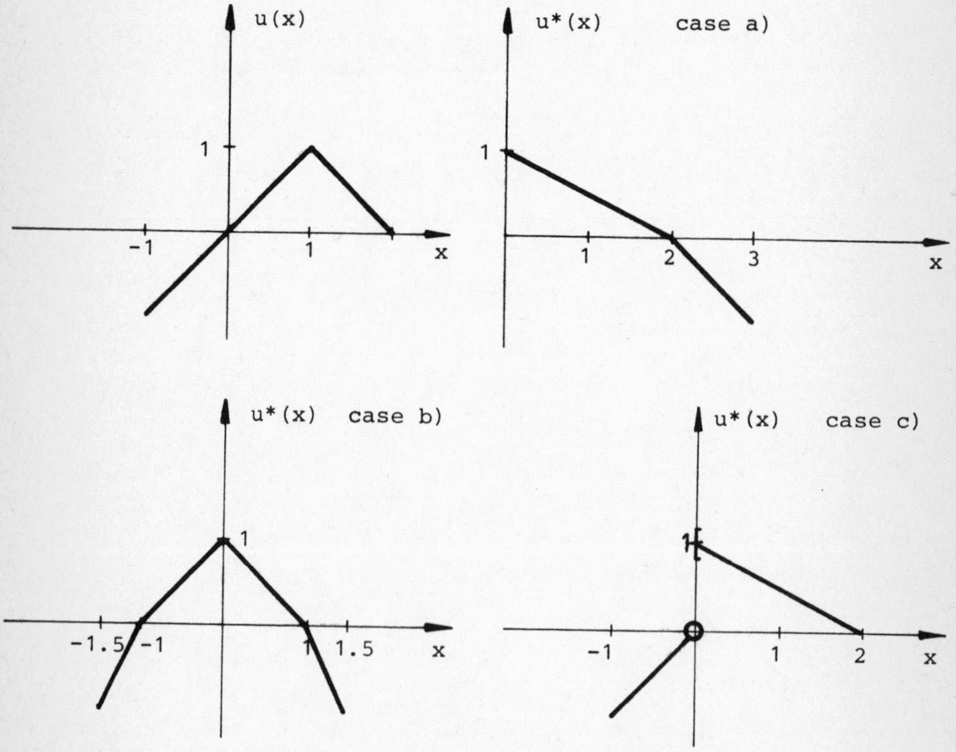

Figure 2.1

Let us now introduce rearrangements of functions of several variables. If $D \subset \mathbb{R}^n$ is a compact set and $n \geq 2$, there is a variety of ways to define D^* .

d) The <u>monotone rearrangement</u> in direction y , also known as linear averaging [131].

For this definition we denote a point $x \in \mathbb{R}^n$ by (x',y) with $x' \in \mathbb{R}^{n-1}$. Furthermore we introduce the notation

$$D(x') = D \cap \{(x',y) | y \in \mathbb{R} \quad \text{for fixed} \quad x' \in \mathbb{R}^{n-1} .$$

The "length" of this one-dimensional set $D(x')$ can be calculated as follows

$$m_1(D(x')) = \int_{\mathbb{R}} \chi_D(x',y)\, dy \quad ,$$

where χ_D is the characteristic function of D , and where the integral is understood as Lebesgue integral. Now we can define

$$D^*(x') := \begin{cases} \{(x',y) \in \mathbb{R}^n | \; 0 \leq y \leq m_1(D(x'))\} & \text{if} \quad D(x') \neq \emptyset \quad , \\ \emptyset & \text{if} \quad D(x') = \emptyset \quad . \end{cases}$$

and

$$D^* := \bigcup_{x' \in D'} D^*(x') \quad ,$$

where $D' \subset \mathbb{R}^{n-1}$ is the set of those $x' \subset \mathbb{R}^{n-1}$ for which $D(x')$ is not empty. We define the <u>monotone (decreasing) rearrangement</u> u^* <u>in direction</u> y of a function u under the following assumption:

$u : \mathbb{R}^n \to \mathbb{R}_0^+$ is a Lipschitz continuous function with compact support $D \subset \bar{\Omega}$, or $\bar{\Omega}$ is a compact set which is convex in y , i.e. for which $\bar{\Omega}(x')$, $x' \in \mathbb{R}^{n-1}$ consists of a single closed interval, and $u : \bar{\Omega} \to \mathbb{R}$ is Lipschitz continuous. (A2.1)

$$u^*(x) := \sup \{c \in \mathbb{R} | \; x \in \Omega_c^*\} \quad \text{for} \quad x \in \bar{\Omega}^* \quad . \qquad (2.1)$$

<u>Remark 2.3</u> Notice that the level sets Ω_c^* of u^* are convex in y and their boundary $\partial\Omega_c^*$ can be described as a function of x' .

e) <u>The starshaped rearrangement</u>

Let us visualize a compact set $D \in \mathbb{R}^n$ in n-dimensional spherical coordinates $(r, \Theta_1, \ldots, \Theta_{n-1})$.

For a point $x = (x_1, \ldots, x_n) \in \mathbb{R}^n$ they are defined by the relations

$$|x| = r ,$$

$$x_1 = r \cos \Theta_1 ,$$

$$\vdots$$

$$x_k = r \sin \Theta_1 \sin \Theta_2 \ldots \sin \Theta_{k-1} \cos \Theta_k , \quad \text{for } k = 2, \ldots, n-1$$

and

$$x_n = r \sin \Theta_1 \sin \Theta_2 \ldots \sin \Theta_{n-2} \sin \Theta_{n-1} , \tag{2.2}$$

where $0 \leq \Theta_k \leq \pi$ for $k = 1, \ldots, n-2$ and $-\pi \leq \Theta_{n-1} \leq \pi$. For typographical reasons let Θ denote the vector of angular coordinates $(\Theta_1, \ldots, \Theta_{n-1})$ and T the $(n-1)$ dimensional hypercube $[0,\pi]^{n-2} \times [-\pi,\pi]$. Then a point $x \in \mathbb{R}^n$ has the representation $x = (r,\Theta)$ where $\Theta \in T$ and $r \in \mathbb{R}_o^+$.

If we formally identify $\Theta \in T$ with $x' \in \mathbb{R}^{n-1}$ from rearrangement d) and r with y, we can transform the set D into a set $\tilde{D}$ which is starshaped with respect to the origin.

<u>Definition:</u>

A set $E \subset \mathbb{R}^n$ is called starshaped with respect to the origin if for every $x \in E$ and every $t \in [0,1]$ the point $t\,x$ is also contained in E.

The corresponding rearrangement u^* of a function would thus lead to a function with starshaped level sets. This is the <u>geometric idea behind starshaped rearrangement</u>. For reasons that will become obvious in §6, we have to measure the distance of ∂D^* to the origin as a weighted average over the characteristic function of D. Let us just mention that the transformation from cartesian the spherical coordinates is singular at the origin and that this circumstance requires some precaution. This is also the reason for the following assumption.

$D \subset \mathbb{R}^n$ is compact and contains a nonempty open ε-neighborhood $U_\varepsilon(o)$ of the origin or D is empty. (A2.2)

If D is empty then D^* is empty by definition. For nonempty D we are going to define D^* as follows.

Let $g : \mathbb{R}^+ \to \mathbb{R}^+$ be a positive and continuous function and let G be a primitive. We define

$$D(\Theta) := \{r \geq \varepsilon \,|\, (r,\Theta') \in D, \Theta' = \Theta\} \qquad \Theta \in T ,$$

$$l(\Theta) := \int_{D(\Theta)} g(r)\, dr , \qquad \Theta \in T ,$$

$$h(\Theta) := l(\Theta) + G(\varepsilon) , \qquad \Theta \in T ,$$

and

$$R(\Theta) := G^{-1}(h(\Theta)) , \qquad \Theta \in T . \tag{2.3}$$

Notice that $R(\Theta)$ does not depent on ε. If $D \subset \mathbb{R}^n$ is compact and contains $U_\varepsilon(o)$, we define

$$D^* := \{x \in \mathbb{R}^n \,|\, 0 \leq |x| \leq R(\Theta)\} . \tag{2.4}$$

In particular we shall be interested in a special class of rearrangements, namely those induced by the family of metrics

$$g(r) = r^{\beta-1} \quad \text{with} \quad \beta \in \mathbb{R} .$$

In addition to assumption (A2.2) let us assume for the moment:

The ray $\{\Theta = \text{const}, r \geq 0\}$ intersects ∂D (transversally) in an odd number $2m + 1$ of points $0 < \varepsilon \leq r_1(\Theta) < r_2(\Theta) < \ldots \leq r_{2m+1}(\Theta)$ with $m = m(\Theta) \in \mathbb{N}_o$. (A2.3)

Then for $\beta \neq 0$ we have

$$h(\Theta) = \frac{1}{\beta}\,(r_1^\beta - r_2^\beta + \ldots + r_{2m+1}^\beta) ,$$

and (2.3a)

$$R(\Theta) = (r_1^\beta - r_2^\beta + \ldots + r_{2m+1}^\beta)^{1/\beta} ,$$

while for $\beta = 0$ we have

$$h(\Theta) = \log r_1 - \log r_2 + \ldots + \log r_{2m+1} ,$$

and (2.3b)

$$R(\Theta) = \frac{r_1\, r_3 \ldots r_{2m+1}}{r_2\, r_4 \ldots r_{2m}} .$$

The case $\beta = n$ is of particular interest, since in this case D and its rearrangement D^* have the same n-dimensional Lebesgue measure. From now on we shall denote this rearrangement by $D^{*(o)}$. Another case of particular interest will be $\beta = n - p$, where $1 \leq p < \infty$. The rearrangement of D under the metric $g(r) = r^{n-p-1}$ will be denoted by $D^{*(p)}$ from now on. The set $\tilde{D}$ on p. 12 above is obtained for $p = n - 1$.

If we want to define the starshaped rearrangement of a function u, we have to make sure that its level sets contain $U_\varepsilon(o)$ or are empty. This will be done by the following assumption.

Let $u : \mathbb{R}^n \to \mathbb{R}_o^+$ be a Lipschitz continuous function with compact support $\bar{\Omega}$. Let $\bar{\Omega}$ contain a nonempty open ε-neighborhood $U_\varepsilon(o)$ of the origin, and let u attain its maximum at each point of $U_\varepsilon(o)$. (A2.4)

Then the (decreasing) starshaped rearrangement u^* of u is defined by

$$u^*(x) := \begin{cases} \sup\{c \in \mathbb{R}_o^+ \mid x \in \Omega_c^*\} & \text{for } x \in \bar{\Omega}^* , \\ 0 & \text{for } x \in \mathbb{R}^n \setminus \bar{\Omega}^* , \end{cases}$$

and in particular,

$$u^{*(p)}(x) := \begin{cases} \sup\{c \in \mathbb{R}_o^+ \mid x \in \Omega_c^{*(p)}\} & \text{for } x \in \bar{\Omega}^{*(p)} , \\ 0 & \text{for } x \in \mathbb{R}^n \setminus \bar{\Omega}^* . \end{cases}$$

f) Steiner symmetrization

This is a multidimensional analogue of the symmetrically decreasing rearrangement b). We use the notation introduced in d) and define the Steiner symmetrization D^* of a compact set $D \subset \mathbb{R}^n$ with respect to the hyperplane $\{y = 0\}$ as follows:

$$D^* = \bigcup_{x' \in D'} D^*(x') ,$$

where

$$
D^*(x') := \begin{cases} \{(x',y) \in \mathbb{R}^n \mid 0 \le |y| \le \frac{1}{2} m_1 \ (D(x'))\} , \\ \qquad\qquad\qquad \text{if } D(x') \neq \emptyset , \\ \emptyset \qquad\qquad\quad \text{if } D(x') = \emptyset . \end{cases} \tag{2.5}
$$

Correspondingly under assumption (A2.1) we define the <u>Steiner symmetrization</u> u^* <u>with respect to the hyperplane</u> $\{y = 0\}$ <u>or the</u> symmetrically decreasing rearrangement u^* of u <u>in the variable</u> y by

$$
u^*(x) := \sup \{c \in \mathbb{R} \mid x \in \Omega_c^*\} \quad \text{for } x \in \bar{\Omega}^* . \tag{2.1}
$$

<u>Remark 2.4</u>

Notice that the level sets of u^* are convex in y and symmetric in y.

<u>Remark 2.5</u>

We want to point out a relation between Steiner symmetrization and starshaped rearrangement in the case $n = 2$ and $g(r) \equiv 1$. Visualize D in polar (r,θ) coordinates. Reflect D across the line $\{r = 0\}$ and consider the union D_r of D and its reflection. Steiner symmetrize D_r with respect to the line $\{r = 0\}$ and restrict the Steiner symmetrized domain D_r^* to the halfplane $\{r \ge 0\}$. This way one obtains the starshaped rearrangement $D^{*(n-1)}$ of D.

g) <u>Schwarz symmetrization</u>

There is more than one way to extend the notion of a symmetric interval of $\mathbb{R}$ to higher dimensions. Under f) the analogue of a symmetric interval was considered to be a family of "parallel" symmetric intervals. If one considers an n-dimensional ball as the generalization of a symmetric interval, then one obtains the notion of Schwarz symmetrization. Hence for a compact set D in $\mathbb{R}^n$ we define the <u>Schwarz symmetrization</u> D^* of D by

$$
D^* := \begin{cases} \text{ball of same n-dim. Lebesgue measure} \\ \text{as } D \text{ with center in the origin ,} \\ \qquad\qquad\qquad \text{if } D \neq \emptyset \ , \\ \\ \emptyset \ , \qquad\qquad \text{if } D = \emptyset \ . \end{cases} \tag{2.6}
$$

For functions satisfying assumption (A2.1) we define

$$
u^*(x) := \sup \{c \in \mathbb{R} \mid x \in \Omega_c^*\} \quad \text{for} \quad x \in \overline{\Omega}^* \ , \tag{2.1}
$$

and call u^* the Schwarz symmetrization of u .

Remark 2.6

This is the most frequently used type of symmetrization. It is named after H.A. Schwarz [167] and is also called spherically symmetric (decreasing) rearrangement [75, 79, 118, 140, 183, 184] or point symmetrization [81, 156, 165]. Spherically symmetric rearrangement should not be confused with "spherical symmetrization" which is defined under h) below. That is why the author of these notes prefers the name Schwarz symmetrization.

Remark 2.7

Polya and Szegö distinguish between Schwarz and point symmetrization. Their definition of "symmetrization of a set with respect to a point" coincides with our definition of Schwarz symmetrization and is commonly referred to a Schwarz symmetrization [19, 118, 140]. "Schwarz symmetrization in the sense of Polya and Szegö" can be applied to the "nepigraph" $D = \{(x,z) \in \mathbb{R}^{n+1} \mid u(x) \geq z \geq 0\}$ of a nonnegative function $u : \overline{\Omega} \to \mathbb{R}_0^+$ with vanishing boundary data. Then the Schwarz symmetrization of D in the sense of Polya and Szegö provides the nepigraph of $u^*(x)$, where $*$ denotes our definition.

Remark 2.8

There is an interesting relationship between Schwarz and Steiner symmetrizations of compact sets in $\mathbb{R}^n$. The Schwarz symmetrization D^* of such a set D can be obtained as a limit of repeated Steiner symmetrizations. This was done for convex sets [156, p. 190; 114, p. 226] but also for nonconvex ones [36].

Remark 2.9

Any result involving Schwarz symmetrization generates open questions of the type: Does the result hold for other rearrangements too? There is a good reason to believe that in general this question has a positive answer. In the next paragraph we shall list a number of properties which do not depend on the type of rearrangement. On the other hand Schwarz symmetrization has some peculiar features which are not shared by other rearrangements. As we shall point out in the applications, a problem involving partial differential equations is often reduced to a (considerably simpler) problem from ordinary differential equations.

Remark 2.10

There are generalizations of Schwarz symmetrizations such as α-symmetrization, in which a plane set D is mapped into a circular sector of angle α, or Schwarz symmetrization on surfaces. We refer the reader to C. Bandle's book for details.

h) Circular and spherical symmetrization

If we visualize a compact domain $D \subset \mathbb{R}^2$ in polar coordinates $r \geq 0$, $\theta \in [-\pi,\pi]$, we can formally Steiner symmetrize D with respect to the angular coordinate θ and obtain a new set D^* which is symmetric in θ and convex in θ. This rearrangement is called circular symmetrization. In cartesian coordinates the geometric effect of this rearrangement is a concentration of D about the positive x_1-axis. Spherical symmetrization is an n-dimensional generalization of this concept, $n \geq 3$: For each fixed $r \geq 0$ the intersection $D(r)$ of D with a sphere of radius r is a subset of $T = [0,\pi]^{n-2} \times [-\pi,\pi]$. Then $D^*(r) = \emptyset$ if $D(r) = \emptyset$ and $D^*(r)$ is the $(n-1)$ dimensional hypercube $[0, \theta_1'] \times [0,\pi]^{n-3} \times$ $\times [-\pi,\pi]$, where θ_1' is chosen in such a way that the $(n-1)$ dimensional Hausdorff measures of $D(r)$ and $D^*(r)$ coincide. In cartesian coordinates $D^*(r)$ is a spherical cap centered around the positive x_1-axis. Finally $D^* := \bigcup_{r \in \mathbb{R}_0^+} D^*(r)$.

Now we can define the circular $(n = 2)$ or spherical symmetrization $(n \geq 3)$ of Lipschitz continuous functions $u : \overline{\Omega} \to \mathbb{R}$, where $\overline{\Omega} \subset \mathbb{R}^n$ is compact, by the usual relation

$$u^*(x) := \sup \{c \in \mathbb{R} \mid x \in \Omega_c^*\} \quad \text{for} \quad x \in \overline{\Omega}^* . \qquad (2.1)$$

i) Radial symmetrization

Suppose $D \subset \mathbb{R}^n$ is a compact set which is starshaped wrt the origin and $\tilde{D} := \frac{1}{2}[D + (-D)] = \{x \in \mathbb{R}^n \mid 2x = y - z \; ; \; y \in D \; , \; z \in D\}$. Then $x \in \tilde{D}$ implies $-x \in \tilde{D}$ and in this sense $\tilde{D}$ has a twofold symmetry. This mapping $D \to \tilde{D}$ is a special case of radial symmetrization and is also known under the name central symmetrization [114]. We shall only define it for the case $n = 2$, and we assume (A2.2).

As in e) let $g : \mathbb{R}^+ \to \mathbb{R}^+$ be a positive and continuous function and let G be a primitive. We define

$$D(\Theta) \;:=\; \{r \geq \varepsilon \mid (r,\Theta') \in D, \Theta = \Theta'\} \quad , \quad \Theta \in [-\pi,\pi] \quad ,$$

$$l(\Theta) \;:=\; \frac{1}{2}\left\{ \int_{D(\Theta)} g(r)\, dr + \int_{D(\Theta+\pi)} g(r)\, dr \right\} \quad ,$$

where $\Theta + \pi \in [-\pi,\pi]$ modulo 2π ,

$$h(\Theta) \;:=\; l(\Theta) + G(\varepsilon) \quad ,$$

$$R(\Theta) \;:=\; G^{-1}(h(\Theta)) \quad , \tag{2.7}$$

and

$$D^* \;:=\; \begin{cases} \{x \in \mathbb{R}^n \mid 0 \leq |x| \leq R(\Theta)\} & \text{if } D \neq \emptyset \; , \\ \emptyset & \text{if } D = \emptyset \; . \end{cases}$$

Notice that $R(\Theta)$ does not depend on ε and that $R(\Theta) = R(\Theta-\pi)$. If one uses $g(r) \equiv 1$ as a weight function this leads to the central symmetrization $\tilde{D}$ above. In analogy to the previous rearrangements the set D^* is called <u>radial symmetrization</u> of D , and for functions u satisfying (A2.4) we define the <u>radial symmetrization</u> u^* as usual by (2.1).

The metric $g(r) = r^{-1}$ was introduced by G. Szegö in [182] for starshaped sets $D \in \mathbb{R}^2$, who called this rearrangement "radial symmetrization".

For those sets we can consider two opposite rays emanating from the origin. Suppose that these rays intersect ∂D transversally in two points P_1 and P_2 . If we measure the distance d_1 and d_2 of P_1 and P_2 to the origin and take their geometric mean, we obtain the distance of ∂D^* to the origin. Here $g(r) = r^{-1}$.

Remark 2.11

Szegö also noted that instead of averaging the distance of ∂D to the origin along two opposite rays one can average over m rays, $2 \leq m \in \mathbb{N}$, which equipartition the unit circle in $\mathbb{R}^2$ and thus generate a starshaped set in $\mathbb{R}^2$ with m-fold symmetry.

Remark 2.12

Heuristically, as m tends to infinity, this process will generate circular level sets. For $G(r) = r$ this is how radial symmetrization can be related to Schwarz symmetrization.

Remark 2.13

M. Marcus considered the case $m = 1$ [130], in which one arrives at starshaped rearrangement e).

Let us summarize that in one space dimension we presented

a) monotone decreasing rearrangement,

b) symmetrically decreasing rearrangement, and

c) quasiconcave rearrangement,

while in higher dimensions there are

d) monotone decreasing rearrangement in direction y,

e) starshaped rearrangement,

f) Steiner symmetrization in direction y,

g) Schwarz symmetrization,

h) circular and spherical symmetrization,

i) radial symmetrization.

Remark 2.14

The rearrangements d) and e) are generalizations of a), while f) g) h) and i) are generalizations of b). e) can also be interpreted as a generalization of c). Some relations between the rearrangements d) through i) are indicated in the diagram below. The numbers refer to corresponding remarks in the text.

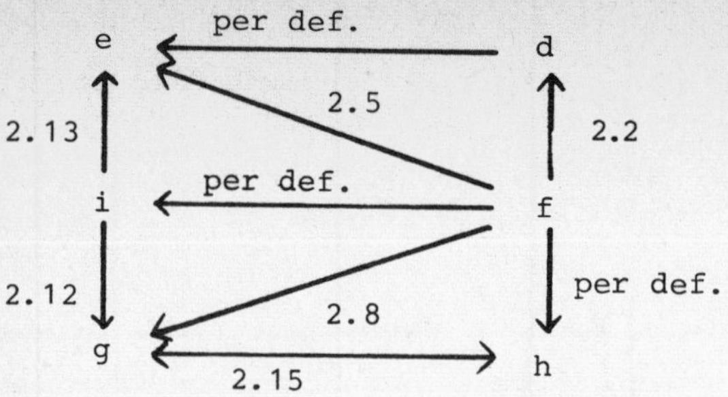

Figure 2.2

Remark 2.15

Spherical and Schwarz symmetrization have a common aspect. Recall that in spherical symmetrization a set $D(r)$ was replaced by a hypercube. Essentially the same happens in Schwarz symmetrization, since a ball in $\mathbb{R}^n$ is nothing but a hypercube (in polar coordinates).

II.2 Common properties of rearrangements

In the preceding paragraph we have defined various kinds of rearrangements for certain classes of functions. We shall from now on refer to these rearrangements a) through i) with the tacit understanding that the assumption (A2.1) and for starshaped c) or radial symmetrization i) also (A2.4) holds. Moreover, in order to allow for a uniform exposition in this paragraph, we shall assume throughout § II.2:

$$\overline{\Omega} = \overline{\Omega}^* \quad . \tag{A2.5}$$

For nonnegative functions with compact support D which does not satisfy $D = D^*$ this is not a serious restriction, since they can be extended by zero. Most of the following properties are already known, but they are scattered in the literature and deserve a uniform treatment.

If $u : \overline{\Omega} \to \mathbb{R}$ is Lipschitz continuous with Lipschitz constant L, then $u^* : \overline{\Omega}^* \to \mathbb{R}$ is Lipschitz continuous with Lipschitz constant L. (L)

Property (L) holds for the rearrangements a) c) d) h), and for the rearrangements b) f) g) only under additional assumptions on u [19, 89, 97, 131]. Without those assumptions and for b) and f) the function u^* is Lipschitz continuous with Lipschitz constant 2L or L' . We refer to § II.4 and § II.7.

If $u : \overline{\Omega} \to \mathbb{R}$ is Lipschitz continuous with Lipschitz constant L , then $u : \overline{\Omega}^* \to \mathbb{R}$ is Lipschitz continuous with Lipschitz constant L' . (L')

A very important property of many rearrangements is the equimeasurability:

The functions u and u^* are equimeasurable, i.e. for $c \in \mathbb{R}$ (E)

$$m_n \{x \in \overline{\Omega} \mid u(x) \geq c\} = m_n \{x \in \overline{\Omega}^* \mid u^*(x) \geq c\} .$$

Property (E) holds for the rearrangements a) b) c) d) f) g) h), and for starshaped rearrangement e) only if the spherical metric $g(r) = r^{n-1}$ is used [79, 88, 97, 156], in which case

$$m_n \{x \in \overline{\Omega} \mid u(x) \geq c\} = m_n \{x \in \overline{\Omega}^{*(o)} \mid u^{*(o)} \geq c\} \text{ for } c \in \mathbb{R} . \quad \text{(E)}$$

Another important feature of rearrangements is that they are order preserving.

$D_1 \subset D_2$ implies $D_1^* \subset D_2^*$ for compact $D_i \subset \mathbb{R}^n$, $i = 1, 2$. (O)

Property (O) holds for all rearrangements a) through i) and implies the following:

The mapping $u \to u^*$ is order preserving, i.e. $u(x) \leq v(x)$ for $x \in \overline{\Omega}$ implies $u^*(x) \leq v^*(x)$ for $x \in \overline{\Omega}^*$. (M1)

Notice that for different rearrangements the domain of definition of "the mapping $u \to u^*$ " is different. For starshaped rearrangement e) for instance $u : \overline{\Omega} \to \mathbb{R}$ has to satisfy (A2.4) and Ω has to satisfy (A2.5). Taking this lack of precision into account, let us list some other properties of the rearrangement mapping $u \to u^*$.

If $c \in \mathbb{R}$ is a constant, then $(u+c)^* = u^* + c$. (M2)

The mapping $u \to u^*$ is positively homogeneous of degree 1 , i.e. for every $t \geq 0$ we have $(tu)^* = tu^*$. (M3)

The mapping $u \to u^*$ is idempotent, i.e. $(u^*)^* = u^*$. (M4)

Properties (M2) (M3) and (M4) hold for all rearrangements a) through i) by definition.

The mapping $u \to u^*$ is nonlinear. (M5)

The mapping $u \to \int_{\overline{\Omega}^*} u^*(x)\,dx$ is additive. (M6)

Property (M6) holds only for equimeasurable rearrangements, but (M5) holds for all a) through i). Let us give a proof of (M5) for the case of monotone decreasing rearrangements. The proof for all other rearrangements is then obvious.

Proof of (M5): Let $\eta(x)$ be a smooth nonnegative function on $\mathbb{R}$ with compact support in $(0,2)$. Let $f_1(x) = \eta(x)$, $f_2(x) = \eta(x-2)$. Then

$$m_1\{(\mathrm{supp}(f_1+f_2))^*\} = 2m_1\{\mathrm{supp}\, f_1\} > m_1\{\mathrm{supp}\, f_1\} =$$

$$= m_1\{\mathrm{supp}(f_1^*+f_2^*)\} .$$

Property (M6) is a consequence of Cavalieri's principle:

For every continuous mapping $F : \mathbb{R}_0^+ \to \mathbb{R}$ and every nonnegative function $u : \overline{\Omega} \to \mathbb{R}_0^+$ (C)

$$\int_{\overline{\Omega}} F(u)\,dx = \int_{\overline{\Omega}^*} F(u^*)\,dx .$$

In fact both integrals in (C) equal $\int_0^\infty F(s)\,d\mu(s)$, where $\mu(s) = m_n(\Omega_s)$, and where the last integral has to be understood as Lebesgue - Stieltjes integral [16, Thm. 1.A; 88, p. 277; 156, pp. 156, 158; 183]. Again let us recall that property (C) holds only for the equimeasurable rearrangements a) b) c) d) f) g) h), and for starshaped rearrangement only in the special case $g(r) = r^{n-1}$. From this point of view one should consider $g(r) = r^{n-1}$ as the most natural metric for starshaped rearrangement. Unfortunately however property (G1) below fails to hold under this metric (cf. § 6, Thm. 2.19).

The equimeasurability of u and u^* has some other convenient consequences. In [58] M. Crandall and L. Tartar showed that (C) and (M1) imply (M7), while (M1) and (M2) imply (M8).

The mapping $u \to u^*$ is nonexpansive in $L^1(\Omega)$, i.e.
$\|u^* - v^*\|_{L^1(\Omega)} \leq \|u - v\|_{L^1(\Omega)}$. (M7)

The mapping $u \to u^*$ is nonexpansive in $L^\infty(\Omega)$. (M8)

Note that (M8) holds without assuming equimeasurability. Moreover they showed that (M7) and (M8) imply

$$\left.\begin{array}{l}\int_{\overline{\Omega}^*} j(|u^*-v^*|)\,dx \leq \int_\Omega j(|u-v|)\,dx \quad \text{for every convex} \\ \text{lower semicontinuous function } j: \mathbb{R}_o^+ \to \mathbb{R}_o^+ \text{ with} \\ j(o) = 0 .\end{array}\right\} \qquad \text{(M9)}$$

Property (M9) had previously been proved by Chiti [50].

In [188] the following property of the L^2-product of u and v is attributed to Hardy and Littlewood

$$\int_{\overline{\Omega}} u(x)\, v(x)\, dx \leq \int_{\Omega^*} u^*(x)\, v^*(x)\, dx \qquad \text{(P1)}$$

for nonnegative functions u,v with compact support in $\overline{\Omega}$.

Proofs of this property are recorded for the rearrangements a) and b) [79, 88], but (P1) holds for all equimeasurable rearrangements. The following Lemma appears to be new:

<u>Lemma 2.1</u> <u>Properties</u> (E) <u>and</u> (O) <u>imply property</u> (P1).

For the proof we extend the notion of rearrangement from Lipschitz continuous to measurable functions. In a first step we show that (P1) holds for characteristic functions. Let $D,E \subset \mathbb{R}^n$ be compact and $u(x) = \chi_D(x)$, $v(x) = \chi_E(x)$. Then the following holds because of (E) and (O):

$$\int_{\mathbb{R}^n} u(x)\, v(x)\, dx = m_n(D \cap E) = m_n(D \cap E)^* \leq m_n(D^* \cap E^*) =$$
$$= \int_{\mathbb{R}^n} u^*(x)\, v^*(x) \quad .$$

In a second step we show that (P1) holds for nonnegative step functions with compact support.

Let

$$u(x) = \sum_{j=1}^{n} a_j \chi_{D_j}(x) \quad \text{and} \quad v(x) = \sum_{k=1}^{m} b_k \chi_{E_k}(x) \quad ,$$

where $a_j, b_k \in \mathbb{R}_o^+$

and where D_j and E_k are compact. Without loss of generality we may assume $D_1 \subset D_2 \subset \ldots \subset D_n$ and $E_1 \subset E_2 \subset \ldots \subset E_m$. Under these special assumptions the rearrangement mapping shows linear behaviour. In fact (compare [79])

$$u^*(x) = \sum_{j=1}^{n} a_j \chi_{D_j^*}(x)$$

and (2.8)

$$v^*(x) = \sum_{k=1}^{m} b_k \chi_{E_k^*}(x) \quad .$$

Then (P1) follows from the first step because of property (M6) and (2.8).

$$\int_{\mathbb{R}^n} u(x)\, v(x)\, dx = \sum_{\substack{j=1,\ldots,n \\ k=1,\ldots,m}} a_j\, b_k\, m_n\, (D_j \cap E_k)$$

$$\leq \sum_{\substack{j=1,\ldots,n \\ k=1,\ldots,m}} a_j\, b_k\, m_n\, (D_j^* \cap E_k^*)$$

$$= \int_{\mathbb{R}^n} \sum_{\substack{j=1,\ldots,n \\ k=1,\ldots,m}} a_j\, b_k\, \chi_{D_j^*}(x)\, \chi_{E_k^*}(x)\, dx$$

$$= \int_{\mathbb{R}^n} u^*(x)\, v^*(x)\, dx \quad .$$

Finally we can prove the assertion by an approximation argument. Any measurable function can be approximated by step functions.

Remark 2.16

The equality in (P1) does not necessarily imply that $u = u^*$ and $v = v^*$. In fact if $u = v$, equality always holds in (P1) because of (C).

Property (P1) also provides a short proof of a special case of (M9): The mapping $u \to u^*$ is nonexpansive in $L^2(\Omega)$. In fact

$$\|u^*-v^*\|^2_{L^2(\overline{\Omega}^*)} = \|u\|^2_{L^2(\Omega)} + \|v\|^2_{L^2(\Omega)} - 2 \int_{\overline{\Omega}^*} u^* v^* \, dx \leq \|u-v\|^2_{L^2(\Omega)} .$$

Another property involving products of functions is the following:

$$\int_{\mathbb{R}^{2n}} f(x)\, g(y)\, h(x-y)\, dx\, dy \leq \int_{\mathbb{R}^{2n}} f^*(x)\, g^*(y)\, h^*(x-y)\, dx\, dy . \tag{P2}$$

Property (P2) is due to F. Riesz [158], and its strict version was discussed in [15, 76, 116]. It is known to hold for the symmetrizations b) and g) [88, 116] and extensions were derived in [36, 116].

What about nonequimeasurable rearrangements, for which property (E) and many of its consequences fail? For starshaped rearrangement under the metric $g(r) = r^{n-1-p}$, $p \geq 1$, the following properties can serve as a substitute for (E) and (C). We refer to § 6 for proofs.

$$m_n(D) = m_n(D^{*(o)}) \geq m_n(D^{*(p)}) \quad \text{for} \quad p \geq 1 . \tag{E'}$$

$$\int_{\Omega} F(u)\, dx \geq \int_{\overline{\Omega}^{*(p)}} F(u^{*(p)})\, dx \quad \text{for} \quad p \geq 1 \tag{C'}$$

for every continuous, monotone nondecreasing mapping $F : \mathbb{R}_o^+ \to \mathbb{R}$ and every nonnegative function $u : \overline{\Omega} \to \mathbb{R}_o^+$ for which $u^{*(p)}$ is defined.

Properties (E') and (C') appear to be new results and have some interesting consequences (cf. § 6). Other nonequimeasurable rearrangements can be found in [103, 110, 111, 151].

Let us now indicate two results on integrals involving the gradient of u. Property (L) says that the L^∞-norm of ∇u does not increase under rearrangements. Therefore the following is not too surprising:

$$\int_{\Omega} |\nabla u|^p \, dx \geq \int_{\Omega^*} |\nabla u^*|^p \, dx \qquad \text{(G1)}$$

for nonnegative functions $u \in W_o^{1,p}(\Omega)$ and for $p > 1$.

Property (G1) is known for the rearrangements a) b) c) d) f) g) h), but for starshaped rearrangement e) only if the metric is induced by $g(r) = r^{n-p-1}$. Notice that for the latter only (E') and not (E) holds [20, 21, 89, 98, 156, 175, 176]. An interesting proof of (G1) for $p = 2$ can be found in [116] and is based on (P2).

Remark 2.17

It was conjectured in [116] and dismissed as hopeless in [156] to prove that the equality sign in (G1) holds only if $u = u^*$. One of the main results of these notes is an investigation of this conjecture for various rearrangements. It will turn out a) that there is a dense subset of the nonnegative functions in $W_o^{1,p}(\Omega)$ for which the conjecture is true ("nice functions") and b) that there is another dense subset, e.g. those piecewise linear functions which are not nice, for which the conjecture is false.

For the equimeasurable rearrangements a) b) c) d) f) g) h) one can prove considerable improvements of (G1), for instance:

$$\int_{\Omega} F(u)\, G(|\nabla u|) \, dx \geq \int_{\Omega^*} F(u^*)\, G(|\nabla u^*|) \, dx \qquad \text{(G2)}$$

for some nonnegative functions $u \in W_o^{1,\infty}(\Omega)$, for convex $G : \mathbb{R}_o^+ \to \mathbb{R}$ and nonnegative continuous $F : \mathbb{R}_o^+ \to \mathbb{R}_o^+$.

We refer to the following paragraphs for more refined statements.

II.3 Monotone decreasing and quasiconcave rearrangement

It is the purpose of this paragraph to familiarize the reader with some basic ideas and additional notation in the context of rearrangement. Essentially the same ideas are used again for other rearrangements, but there they might be obscured by technical details.

Throughout this paragraph $*$ denotes the monotone decreasing rearrangement a) and as in the definition in § II.1 we assume that $u : \mathbb{R} \to \mathbb{R}_o^+$ is a Lipschitz continuous function with compact support D in $\bar{\Omega}$ or that $\bar{\Omega}$ is a compact interval and $u : \bar{\Omega} \to \mathbb{R}$ is Lipschitz continuous. According to Remark 2.1 the results can be immediately extended to quasiconcave rearrangement c).

Let us first compare our definition of monotone decreasing rearrangement with the one given by Hardy, Littlewood and Polya, who introduced the distribuction function $\rho(c) := m_1(\Omega_c)$ for $c \in \mathbb{R}$. This function is monotone nonincreasing because $\Omega_b \supset \Omega_c$ for $b < c$, and continuous from the left [159, p. 61].

Moreover the function $\rho(c)$ is monotone decreasing on $\left[\min_{\Omega} u \ , \max_{\Omega} u\right]$. (2.9)

In fact if ρ were not decreasing, then $\rho(c_1) = \rho(c_2)$ for two real numbers $c_1 < c_2$, $c_1, c_2 \in \left(\min u \ , \max u\right)$. Hence the set $A := \{x \in \bar{\Omega} | \ c_2 > u(x) \geq c_1\}$ has Lebesgue measure zero. But since u is continuous there exists an x_o such that $c_1 < u(x_o) < c_2$ and A contains a δ-neighborhood $U_\delta(x_o)$ for sufficiently small positive δ , a contradiction.

The distribution function ρ has discontinuities in those points $c \in \left[\min_{\Omega} u \ , \max_{\Omega} u\right]$, for which $u(x) = c$ on a set of positive measure. This is illustrated in Fig. 2.3.

Now let $\omega := m_1(\bar{\Omega})$ and let $\bar{\rho} : \left[\min_{\Omega} u \ , \max_{\Omega} u\right] \to 2^{[o,\omega]}$ denote the following multivalued extension of ρ :

$$\bar{\rho}(c) = \begin{cases} \{\rho(c)\} & \text{for } c = \max_{\Omega} u \ , \\ [\rho(c+o), \rho(c-o)] & \text{for } c \in \left(\min_{\Omega} u \ , \max_{\Omega} u\right) , \\ \{\rho(c)\} & \text{for } c = \min_{\Omega} u \ . \end{cases}$$

Then the inclusion $v \in \bar{\rho}(c)$ has because of (2.9) a unique solution c for each $v \in [0,\omega]$, i.e. the inverse function $v \in \bar{\rho}^{-1}\{v\}$ exists and will be called $\tilde{u}^*(v)$. By construction

$$\tilde{u}^*(v) \;:=\; \sup\left\{c \in \left[\min_{\Omega} u \;,\; \max_{\Omega} u\right] \mid v \le \rho(c)\right\} \quad \text{for} \quad v \in [0,\omega]$$

or equivalently

$$\tilde{u}^*(v) \;:=\; \sup\,\{c \in \mathbb{R} \mid v \in \Omega_c^*\} \qquad\qquad \text{for} \quad v \in [0,\omega] \quad .$$

Therefore $\tilde{u}^* = u^*$. An example of a function u together with ρ and u^* is drawn below in Fig. 2.3.

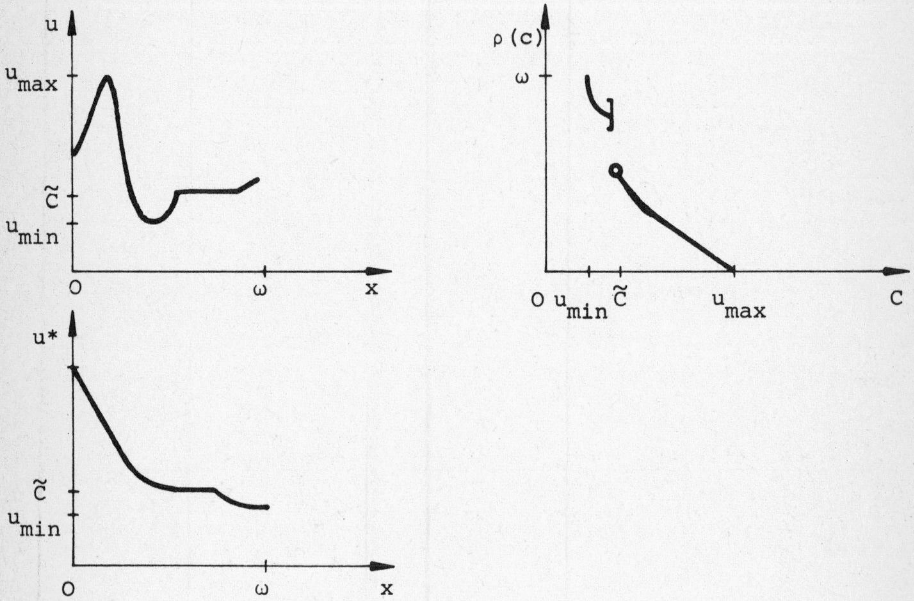

Fig. 2.3

In the following lemma we state that the level sets of u^* are in fact the rearrangements of the level sets of u .

Lemma 2.2 $\Omega_c^* = \{x \in \bar{\Omega}^* \mid u^*(x) \ge c\}$ for $c \in \mathbb{R}$.

For $c \le \min_{\Omega} u$ and $c > \max_{\Omega} u$ the proof is trivial. For $\min_{\Omega} u < c < \max_{\Omega} u$ by definition $u^*(x) \ge c$ if and only if $x \in \Omega_b^*$ for all $b < c$, i.e. $x \in \bigcap_{b<c} \Omega_b^*$. But trivially we have $\Omega_c^* \subset \bigcap_{b<c} \Omega_b^*$, and it remains to show $\Omega_c^* \supset \bigcap_{b<c} \Omega_b^*$ or equivalently $m_1(\Omega_c) \ge \lim_{\substack{b<c \\ b\to c}} m_1(\Omega_b)$. The latter is a standard result from measure theory [159, p. 61].

For the proof of the following lemma we make the additional assumption:

$$\overline{\Omega} = [0,\omega] \text{ , where } 0 < \omega \in \mathbb{R} \text{ .} \tag{A2.5a}$$

Remark 2.18

This assumption can be made without loss of generality. If $\overline{\Omega}$ is simply connected and its left endpoint does not coincide with zero, then we can translate the origin. If $\overline{\Omega}$ is not simply connected, then we can extend the function u by zero and replace $\overline{\Omega}$ by its closed convex hull.

Notice that if u does not have zero boundary values and $\overline{\Omega}$ is not simply connected, the following Lemma 2.3 cannot be true. A counterexample is $u(x) = x$, $x \in [0,1] \cup [2,3]$.

Lemma 2.3 Property (L) holds, i.e. if u is Lipschitz continuous with Lipschitz constant L , then u^* is Lipschitz continuous with the same constant.

For the proof we assume without loss of generality (A2.5a). First we observe that in addition to (2.9) the distribution function ρ cannot be "flat". In fact it always has negative slope bounded away from zero by $-\frac{1}{L}$.

$$\rho(c_1) - \rho(c_2) \leq -\frac{1}{L}(c_1-c_2) \quad \text{for} \quad c_1 \neq c_2,\ c_i \in \left[\min_{\Omega} u \ , \max_{\Omega} u\right] . \tag{2.10}$$

This can be seen as follows. If $c_1 > c_2$ and $c_i \in$ range u then there exists an x_1 and x_2 such that $u(x_i) = c_i$, $i = 1,2$ and such that $|x_1 - x_2| = d(\Omega_{c_1}, \partial\Omega_{c_2})$. By assumption $c_1 - c_2 \leq L|x_1 - x_2|$, and $x_i \in \Omega_{c_i}$, $i = 1, 2$. Therefore

$$c_1 - c_2 \leq L\, m_1 \{\Omega_{c_2} \setminus \Omega_{c_1}\} = L\,(m_1(\Omega_{c_2}) - m_1(\Omega_{c_1})) = L\,(\rho(c_2) - \rho(c_1)) ,$$

which proves (2.10).

Therefore the inverse function ρ^{-1} : range $\rho \cap [0,\omega] \to [\min u, \max u]$ has slope greater than or equal to $-L$, i.e. for $x > y$, $x,y \in [0,\omega] \cap$ range ρ we have

$$y - x < -\frac{1}{L}[\rho^{-1}(x) - \rho^{-1}(y)] \quad \text{or} \quad \rho^{-1}(x) - \rho^{-1}(y) \leq L(x-y) \quad .$$

The graph of u^* consists of the graph of ρ^{-1} and horizontal intervals. This and the monotonicity of u^* imply its Lipschitz continuity.

Lemma 2.3 implies that u^* is almost everywhere differentiable. Therefore the natural question arises whether u^* inherits higher differentiability properties from u. For sufficiently well behaved functions u this question can be answered in the positive by explicit calculation of derivatives of u^*.

Definition:

In this paragraph we call a function $u : \overline{\Omega} \to \mathbb{R}$ simple if and only if $\overline{\Omega} = \overline{\Omega}^*$, if $u \in C(\overline{\Omega})$ and if u is piecewise linear (in the sense of affine).

We call a function $u : \overline{\Omega} \to \mathbb{R}$ nice if and only if it is simple and $\frac{du}{dx} \neq 0$ in the interior of each subinterval.

We call a function $u : \overline{\Omega} \to \mathbb{R}$ smooth if and only if $u \in C^1(\overline{\Omega})$, if $\overline{\Omega} = \overline{\Omega}^*$ and if u has the following properties:

i) the set of points $\{x \in \Omega |\ u(x) = c\}$ is finite for each $c \in \left(\min_{\Omega} u , \max_{\Omega} u\right)$,

ii) the set of points $\left\{x \in \Omega \,\middle|\, \frac{du}{dx} = 0\right\}$ is finite.

Notice that simple and smooth functions have a convenient feature in common. We can calculate $\rho(c) = m_1(\Omega_c)$ explicitly for $c \in \left(\min_{\Omega} u , \max_{\Omega} u\right)$ by adding up a finite number of intervals.

The main purpose of the following lemma is to familiarize the reader with some notation which is needed in the proof of Lemma 2.6.

Lemma 2.4 Suppose (A2.5a) holds, i.e. $\overline{\Omega} = [0,\omega]$. Let $u : \overline{\Omega} \to \mathbb{R}$ be simple or smooth and let the function u attain its maximum in the origin. Then there exist two open subsets O_1 and O_2 of Ω such that u^* is constant in O_1 and u^* is differentiable and $\left|\frac{du^*}{dx}\right| > 0$ in O_2. Furthermore $O_1 \cup O_2$ is dense in Ω and $\Omega \setminus (O_1 \cup O_2)$ consists of finitely many points.

Proof: If u is simple, then the endpoints of the intervals in which u is linear induce a finite partition of $\bar{\Omega}$. For smooth functions such a partition is generated by the finite set of points $\left\{x \in \Omega \,\middle|\, \frac{du}{dx} = 0\right\}$. Without loss of generality let there be M points which partition $\bar{\Omega}$ in the above manner. Let $\{a_i\}_{i=1,\dots,M}$ be the increasingly ordered values assumed by u at these M points. If the set $\bigcup_{i=1}^{M} \{x \in \bar{\Omega} \mid u(x) = a_i\}$ has positive measure, i.e. if u is constant on certain subintervals, then $\frac{du}{dx} = 0$ a.e. on these intervals. Analogously $\frac{du^*}{dx} = 0$ a.e. on the intervals $\bigcup_{i=1}^{M} \{x \in \bar{\Omega} \mid u^*(x) = a_i\}$ the interior of which we call O_1. Now O_2 will be the open set $(0,\omega) \setminus \bigcup_{i=1}^{M} \{x \in \bar{\Omega} \mid u^*(x) = a_i\}$. Notice that $\Omega \setminus \{O_1 \cup O_2\}$ consists of finitely many points, so that $O_1 \cup O_2$ is dense in Ω and $\Omega \setminus (O_1 \cup O_2)$ has measure zero. It remains to show that u^* is differentiable in O_2 and that $\left|\frac{du^*}{dx}\right| > 0$ there.

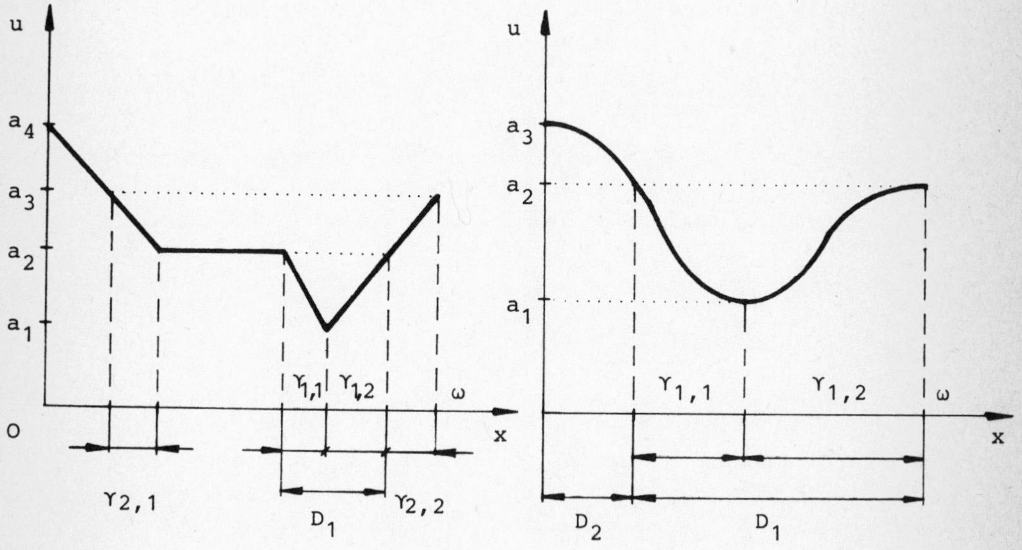

Fig. 2.4

To this end define $D_i = \{x \in \bar{\Omega} \mid a_i < u(x) < a_{i+1}\}$ and $D_i^* = \{x \in \bar{\Omega} \mid a_i < u^*(x) < a_{i+1}\}$ for $i = 1, \dots, M-1$. Now fix i and decompose D_i into a finite number of intervals $\{\gamma_{i,j}\}_{j=1,\dots,N(i)}$, in each of which u is either linear (in the sense of affine) and nonconstant, if u is simple, or in which $\frac{du}{dx}$

does not vanish if u is smooth. For each $\lambda \in (a_i, a_{i+1})$ and $j = 1, \ldots, N(i)$ we denote by $\rho_j(\lambda)$ the unique value of ρ in $\gamma_{i,j}$ for which $u(\rho) = \lambda$. The function $u^*(x)$ is monotone decreasing in D_i^*. We denote by $\rho^*(\lambda)$ the value of ρ in D_i^* for which $u^*(\rho^*) = \lambda$. Now we make the additional convention that for each $i = 1, \ldots, M-1$ the intervals $\gamma_{i,j}$ are ordered by their distance from the origin with $\gamma_{i,1}$ being closest to zero, so that by definition of u^*

$$\text{sign} \frac{du}{dx} = (-1)^j \quad \text{in} \quad \gamma_{i,j} \tag{2.11}$$

and

$$\rho^*(\lambda) = \begin{cases} \sum_{j=1}^{N} (-1)^{j+1} \rho_j(\lambda) & \text{if } N \text{ is odd}, \\ \\ \sum_{j=1}^{N} (-1)^{j+1} \rho_j(\lambda) + \omega & \text{if } N \text{ is even}. \end{cases} \tag{2.12}$$

Notice that at this place we use the assumption that u attains its maximum at the origin. In the interior of each interval $\gamma_{i,j}$ the function $\rho_j(\lambda)$ is differentiable, since $\frac{du}{dx}$ does not vanish there and we have

$$\frac{du}{dx} = 1/\left(\frac{d\rho_j}{d\lambda}\right) \quad \text{in} \quad \gamma_{i,j} \quad . \tag{2.13}$$

Since each of the functions ρ_j is differentiable with respect to λ, so is ρ^* because of (2.12) for $\lambda \in (a_i, a_{i+1})$. Recall that $\infty > \left|\frac{d\rho^*}{d\lambda}\right| \geq \frac{1}{L} > 0$ for $\lambda \in (a_i, a_{i+1})$, where L is the uniform Lipschitz constant for u, and apply the inverse function theorem to see that u^* is differentiable in O_2 and that

$$\frac{du^*}{dx} = 1/\left(\frac{d\rho^*}{d\lambda}\right) \tag{2.14}$$

does not vanish in O_2. This concludes the proof of Lemma 2.4. Another proof was given in [63].

Remark 2.19

In more than one space dimension, e.g. for starshaped rearrangement, the function ρ_j and also ω will depend on additional variables, e.g. Θ . Then the case that N is even in (2.12) will have to be excluded by imposing additional assumptions on u .

Remark 2.20

The assumption of Lemma 2.4 that u attains its maximum in the origin is not necessary and was only made to simplify the proof. If this assumption is given up, then (2.11) has to be replaced by

$$\operatorname{sign} \frac{du}{dx}(\rho_j) = (-1)^{j+1} \operatorname{sign} \frac{du}{dx}(\rho_1) ,$$

and the remaining changes are left as an exercise for the reader.

Remark 2.21

If u is not only simple, but nice, or if u is smooth then O_1 is empty.

Lemma 2.5 *Suppose in addition to the assumption of Lemma 2.4 that* u *is smooth and* $u \in C^k(\Omega)$. *Then* $u^* \in C^k(O_2)$.

The proof follows from the observation that u^* is the inverse function of ρ^* . In O_2 the derivative $\frac{d\rho^*}{d\lambda}$ does not vanish. Hence ρ^* is k-times differentiable because of (2.12) and the implicit function theorem.

Now we are going to derive property (G2). A different proof of the first part of the following lemma was given in [64].

Lemma 2.6 *Let* $u : \overline{\Omega} \to \mathbb{R}$ *be simple or smooth and let the function* u *attain its maximum at the origin. Let* $F : \mathbb{R} \to \mathbb{R}$ *be nonnegative and continuous and let* $G : \mathbb{R}_o^+ \to \mathbb{R}$ *be monotone nondecreasing and convex.*

i) *Then the inequality*

$$\int_\Omega F(u)\, G\left(\left|\frac{du}{dx}\right|\right) dx \geq \int_{\Omega^*} F(u^*)\, G\left(\left|\frac{du^*}{dx}\right|\right) dx \qquad \text{(G2a)}$$

holds.

ii) <u>If in addition</u> F <u>is positive and</u> G <u>monotone increasing, then equality in</u> (G2a) <u>holds only if</u> $u = u^*$.

<u>Proof:</u> We use the notations introduced in the proof of Lemma 2.4. We observe $m_1\,(x \in \overline{\Omega} \mid u(x) = a_i) = m_1\,(x \in \overline{\Omega}^* \mid u^*(x) = a_i)$. Then it suffices to show

$$\int_{D_i} F(u(x))\, G\left(\left|\frac{du}{dx}\right|\right) dx \geq \int_{D_i} F(u^*(x))\, G\left(\left|\frac{du^*}{dx}\right|\right) dx \qquad \text{for } i = 1, \ldots, M-1 \; , \tag{2.15}$$

since we can disregard sets of positive measure on which u^* or u might be constant, if u is simple. Let us now fix i and introduce λ as the variable of integration in inequality (2.15). This is permitted because of Lemma 2.4. Hence inequality (2.15) can be rewritten as follows

$$\sum_{j=1}^{N} \int_{\gamma_{i,j}} F(u)\, G\left(\left|\frac{du}{dx}\right|\right) dx = \sum_{j=1}^{N} \int_{a_i}^{a_{i+1}} F(\lambda)\, G\left(\left|\frac{d\rho_j}{d\lambda}\right|^{-1}\right) \left|\frac{d\rho_j}{d\lambda}\right| d\lambda$$

$$\geq \int_{a_i}^{a_{i+1}} F(\lambda)\, G\left(\left|\sum_{j=1}^{N} (-1)^{j+1} \frac{d\rho_j}{d\lambda}\right|^{-1}\right) \left|\sum_{j=1}^{N} (-1)^{j+1} \frac{d\rho_j}{d\lambda}\right| d\lambda \; . \tag{2.16}$$

Using (2.11) (2.12) (2.13) one can easily see that

$$\left|\sum_{j=1}^{N} (-1)^{j+1} \frac{d\rho_j}{d\lambda}\right| = \sum_{j=1}^{N} \left|\frac{d\rho_j}{d\lambda}\right| \; ,$$

so that the relation (2.16) is equivalent to

$$\int_{a_i}^{a_{i+1}} \sum_{j=1}^{N} F(\lambda)\, G\left(\left|\frac{d\rho_j}{d\lambda}\right|^{-1}\right) \left|\frac{d\rho_j}{d\lambda}\right| d\lambda \geq$$

$$\geq \int_{a_i}^{a_{i+1}} F(\lambda)\, G\left(\left[\sum_{k=1}^{N} \left|\frac{d\rho_k}{d\lambda}\right|\right]^{-1}\right) \sum_{k=1}^{N} \left|\frac{d\rho_k}{d\lambda}\right| d\lambda \; . \tag{2.17}$$

In order to show (2.17) we have to verify

$$\sum_{j=1}^{N} \alpha_j\, G\left(\left|\frac{d\rho_j}{d\lambda}\right|^{-1}\right) \geq G\left(\left[\sum_{k=1}^{N} \left|\frac{d\rho_k}{d\lambda}\right|\right]^{-1}\right) \; , \tag{2.18}$$

where

$$\alpha_j \;:=\; \left|\frac{d\rho_j}{d\lambda}\right| \left[\sum_{k=1}^{N} \left|\frac{d\rho_k}{d\lambda}\right|\right]^{-1} \quad \text{and} \quad \sum_{j=1}^{N} \alpha_j \;=\; 1 \;.$$

But (2.18) is a consequence of the convexity and monotonicity of G. In fact we have

$$\sum_{j=1}^{N} \alpha_j \, G\left(\left|\frac{d\rho_j}{d\lambda}\right|^{-1}\right) \;\geq\; G\left(\sum_{j=1}^{N} \alpha_j \left|\frac{d\rho_j}{d\lambda}\right|^{-1}\right)$$

$$= \; G\left(N \left[\sum_{k=1}^{N} \left|\frac{d\rho_k}{d\lambda}\right|\right]^{-1}\right)$$

$$\geq \; G\left(\left[\sum_{k=1}^{N} \left|\frac{d\rho_k}{d\lambda}\right|\right]^{-1}\right) .$$

This proves (G2a). To see that the equality sign holds only if $u = u^*$, we suppose that $u \neq u^*$, i.e. that u is not monotone nonincreasing in x. Then there exists an $i \in \{1,\ldots,M-1\}$ such that $N(i) \geq 2$ on D_i and such that the strict inequality holds in (2.17).

Remark 2.22

As in Lemma 2.4 the assumption that u attains its maximum in the origin is not necessary. If one gives up this assumption, however, the second statement of Lemma 2.6 has to be changed to: Equality in (G2a) holds only if u is monotone.

Remark 2.23a

Since simple functions are dense in $W^{1,p}(\Omega)$, it is possible to extend Lemma 2.6 i) to functions in $W^{1,p}(\Omega)$, $\infty > p > 1$, and to obtain the inequality

$$\int_{\Omega^*} \left|\frac{du^*}{dx}\right|^p dx \;\leq\; \int_{\Omega} \left|\frac{du}{dx}\right|^p dx \qquad \text{(G1a)}$$

for all functions $0 \leq u \in W^{1,p}(\Omega)$ and for $\infty > p > 1$ (cf. also Corollary 2.10). The case $p = \infty$ is treated in Lemma 2.3.

The equality sign in (G1a) is discussed in [62] for absolutely continuous functions. To this end a nonmonotone function u is approximated by simple functions u_n, and then one has to show that the "defect"

$$d(n) = \left\| \frac{du_n}{dx} \right\|_{L^p(\Omega)} - \left\| \frac{du_n^*}{dx} \right\|_{L^p(\Omega)}$$

is bounded below by a positive constant which is independent of n.

This last step, however, was left out as "easy" in [62, p. 469]. At least for $p = 1$ though and for absolutely continuous functions u one can show that the equality sign in (G1a) implies that u is monotone by looking at the total variation of u [87].

An improvement of (G1a) was derived in [63], namely

$$\int_{\Omega^*} \left|\frac{du}{dx}\right|^p dx \leq \int_{\Omega} \left\{\frac{1}{N(\lambda)} \left|\frac{du}{dx}\right|\right\}^p dx ,$$

where $N(\lambda)$ is the number of times that the constant function λ crosses the graph of u, which appears in (2.12).

In [160] one can also find inequality (G1a) for $p < 1$. In [62, Thm. 7.1] it is claimed that (G1a) remains true if $\frac{d}{dx}$ is replaced by $\frac{d^2}{dx^2}$. This extension is false as one can see from the counter-example $u(x) = 8 + 2x^2 - x^4$, $0 \leq x \leq 3$. Some other results of Duff and Ryff are not undisputed, see [147].

Remark 2.23b

Recently it was shown by I. Klemes [109] that the BMO norm of u decreases under monotone rearrangement.

Open problem:

Find optimal assumptions on the regularity of u, under which Lemma 2.6 ii) holds.

II.4 Symmetric decreasing rearrangement

While § II.3 was intended as a preparation for § II.7 on starshaped rearrangement, this paragraph is meant to prepare the reader for the problems and potential properties of Steiner symmetrization, which is treated in § II.5.

Unless otherwise stated we assume throughout this paragraph that u^* denotes the symmetrically decreasing rearrangement of u, that

$$\bar{\Omega} = [-\omega,\omega] = \bar{\Omega}^*, \quad \omega \in \mathbb{R}^+ \tag{A2.5b}$$

holds, and that $u : \bar{\Omega} \to \mathbb{R}$ is a Lipschitz continuous function. Remark 2.18 applies correspondingly, i.e. assumption (A2.5b) can be made without loss of generality.

The following example shows that in general property (L) fails for this type of rearrangement.

Example 2.2

Let $\bar{\Omega} = [-1,1]$ and $u(x) = x$. Then $u^*(x) = 1-2\,|x|$ on $\bar{\Omega}$ and u^* is Lipschitz continuous with Lipschitz constant 2.

Lemma 2.7 *Suppose (A2.5b) holds. If $u : \bar{\Omega} \to \mathbb{R}$ is Lipschitz continuous with Lipschitz constant L, then $u^* : \bar{\Omega}^* \to \mathbb{R}$ is Lipschitz continuous with Lipschitz constant $2L$.*

The proof of this lemma is a straightforward modification of the proof of Lemma 2.3. The factor 2 comes into play since u^* restricted to $[0,\omega]$ is essentially the inverse of $\frac{1}{2}\rho$.

Under additional assumptions one can prove improvements of Lemma 2.7. For instance if we assume

$$u : \bar{\Omega} \to \mathbb{R}_0^+ \quad \text{and} \quad u = 0 \text{ on } \partial\Omega , \tag{A2.6}$$

then we can prove the following.

Lemma 2.8 *Suppose (A2.5b) and (A2.6) hold and $u : \bar{\Omega} \to \mathbb{R}_0^+$ is Lipschitz continuous with Lipschitz constant L, then $u^* : \bar{\Omega}^* \to \mathbb{R}_0^+$ is Lipschitz continuous with the same constant.*

For the proof we continue u and u^* outside their domain of definition by zero and, abusing notation, denote the extended functions by u and u^*.

We intend to show $|u^*(y) - u^*(x)| \leq L\,|x-y|$ for every $x,y \in \overline{\Omega}^*$. Without loss of generality we may assume $u^*(x) < u^*(y)$ and $x,y \in [0,\omega]$. Then we have to show

$$u^*(x) \geq u^*(y) - L\,|x-y| \quad . \tag{2.19}$$

To this end we define $E_y := \{z \in \overline{\Omega}^* |\ u(z) \geq u^*(y)\}$. If we consider $u^*(y)$ to be a given constant, the analogue of Lemma 2.2 yields

$$E_y^* = \left\{z \in \overline{\Omega}^* |\ u^*(z) \geq u^*(y)\right\} ,$$

and obviously $y \in E_y^*$. Furthermore we define $E_x := \{z \in \mathbb{R} |\ d(z,E_y) \leq |x-y|\}$.

By definition for every $z \in E_x$ there exists a $\tilde{y} \in E_y$ such that $u(\tilde{y}) \geq u^*(y)$ and $|\tilde{y}-z| \leq |x-y|$. Hence $u(z) - u^*(y) = u(z) - u(\tilde{y}) + u(\tilde{y}) - u^*(y) \geq u(z) - u(\tilde{y}) \geq - L\,|z-\tilde{y}| \geq - L\,|x-y|$, and $E_x \subset \{z \in \mathbb{R} |\ u(z) \geq u^*(y) - L\,|x-y|\}$.

Now we recall that rearrangements are order preserving, so that

$$E_x^* \subset \{z \in \mathbb{R} |\ u(z) \geq u^*(y) - L\,|x-y|\}^*$$

$$= \{z \in \mathbb{R} |\ u^*(z) \geq u^*(y) - L\,|x-y|\} \quad .$$

But the "length" $m_1(E_x^*)$ of E_x exceeds the "length" of E_y by at least $2\,|x-y|$ and $y \in E_y^*$. Therefore $x \in E_x^*$, which proves (2.19). The idea for this proof was extracted from [131].

<u>Remark 2.24</u>

Another way to prove Lemma 2.8 would be by means of property (G2b) which is derived in Corollary 2.10 below, by setting $G(t) = t^p$ and sending p to infinity. This shows that assumption (A2.6) is not optimal. It can be replaced for instance by the periodicity assumption

$$u : [-\omega,\omega] \to \mathbb{R} \quad , \quad u(\omega) = u(-\omega) \quad . \tag{A2.7}$$

The periodicity assumption can be reduced to (A2.6) in the following way. Extend u periodically to $\mathbb{R}$ and call the extension u. After possibly adding a constant to u and after translating the origin one can make sure that u restricted to $\overline{\Omega}$ satisfies (A2.6). On the other hand (A2.6) is a special case of (A2.7).

As in § 3 we can define simple and nice functions, see p. 30.

Remark 2.25

Nice functions are dense in $W^{1,p}(\Omega)$, $1 < p < \infty$, $C^\infty(\Omega)$ is dense in $W^{1,p}(\Omega)$ and every C^∞-function can be approximated through piecewise linear interpolation by simple functions in the $W^{1,p}$ norm. Finally any simple function can be approximated by nice functions in the following way. If u is simple but not nice and $\overline{\Omega} = [-\omega,\omega]$, determine the positive number

$$\varepsilon_o := \min \left\{ \left|\frac{du}{dy}\right| , y \in \overline{\Omega} , \frac{du}{dx}(y) \text{ exists and does not vanish} \right\} .$$

For $\varepsilon < \varepsilon_o$ consider the auxiliary function $v_\varepsilon(y) = \varepsilon(1-|y|/\omega)$ for $y \in \overline{\Omega}$. Then $u_\varepsilon(y) := u(y) + v_\varepsilon(y)$ is a nice function and u_ε converges to u in $W^{1,p}(\Omega)$ as ε tends to zero.

Theorem 2.9

Let the function u be nice or analytic on $\overline{\Omega}$, suppose that (A2.5b) holds and at least one of (A2.6) or (A2.7). Let F : range $u \to \mathbb{R}_o^+$ be nonnegative and continuous and let $G : \mathbb{R}_o^+ \to \mathbb{R}$ be monotone nondecreasing and convex:

i) Then the following inequality holds

$$\int_\Omega F(u)\, G\left(\left|\frac{du}{dx}\right|\right) dy \geq \int_{\Omega^*} F(u^*)\, G\left(\left|\frac{du^*}{dx}\right|\right) dy . \qquad \text{(G2b)}$$

ii) If moreover F is positive and G monotone increasing and strictly convex, then equality holds in (G2b) if and only if $u = u^*$ modulo translation. If (A2.6) holds, equality holds in (G2b) if and only if $u = u^*$.

Remark 2.26

We call a function u analytic on $\overline{\Omega}$ if u has an analytic continuation across $\partial\Omega$. It is possible to extend this theorem to functions which are Lipschitz continuous on $\overline{\Omega}$ and analytic in the domain Ω, such as $(\omega - x)^2 \cdot \sin \frac{1}{(x-\omega)}$. In this case one has to deal with a possibly countable partition of $\overline{\Omega}$ into sets D_i . Nevertheless for all but a finite number of λ's the constant function λ intersects the graph of u only finitely often.

Remark 2.27

The assumptions (A2.6) or (A2.7) are not optimal. What is needed for the proof is merely the condition (A2.8):

For almost every $\lambda \in$ range u the graph of the constant function λ intersects the graph of u at least twice, in short $N(\lambda) \geq 2$ a.e. λ . (A2.8)

If one assumes (A2.8) only, one has to distinguish a few more cases in the representation (2.21) below, but otherwise the proof is the same. If we give up (A2.8) alltogether, Theorem 2.9 cannot be true as Example 2.2 shows.

Proof of Theorem 2.9

Since u^* is Lipschitz continuous, the right hand side of (G2b) makes sense. If u is nice, the endpoints of the intervals in which u is affine form a finite partition of $\overline{\Omega}$. If u is analytic on $\overline{\Omega}$ then either $u \equiv \text{const}$, in which case there is nothing to prove, or $\frac{du}{dy}$ has only finitely many zeros in $\overline{\Omega}$ which form a partition of $\overline{\Omega}$. Let there be M such points and let $0 = a_1 \leq a_2 \leq \ldots \leq a_M$ be the values assumed by u at these M points. Define $D_i := \{y \in \Omega | \; a_i < u(y) < a_{i+1}\}$ and $D_i^* := \{y \in \Omega | \; a_i < u^*(y) < a_{i+1}\}$ for $i = 1, \ldots, M-1$. It suffices to integrate over the union of the sets D_i (D_i^*) on the left (right) hand side of (G2b). Fix i and decompose D_i into a finite number of intervals $\{\gamma_{i,j}\}_{j=1,\ldots,N(i)}$ in each of which $\frac{du}{dy}$ exists and does not vanish. For each $\lambda \in (a_i, a_{i+1})$ denote by $y_j(\lambda)$ the unique value of y in $\gamma_{i,j}$ for which $u(y_j) = \lambda$. The function u^* is monotone decreasing on $[0,\omega]$. We denote by $y^*(\lambda)$ the nonnegative value of y in D_i^* for which $u^*(y^*) = \lambda$. We make the convention that for each i the intervals $\{\gamma_{i,j}\}_{j=1,\ldots,N}$ are ordered by their distance from $-\omega$, so that by definition

$$\text{sign} \frac{du}{dy}(y_j) = (-1)^{j+1} \text{sign} \frac{du}{dy}(y_1) \quad \text{in } \gamma_{i,j} \qquad (2.20)$$

and

$$y^*(\lambda) = \begin{cases} \frac{1}{2} \sum_{j=1}^{N} (-1)^j \; y_j(\lambda) & \text{if} \quad \text{sign} \frac{du}{dy} (y_1) = +1 \quad \text{in} \quad (a_i, a_{i+1}) \\ \frac{1}{2} \sum_{j=1}^{N} (-1)^{j+1} \; y_j(\lambda) + \omega & \text{if} \quad \text{sign} \frac{du}{dy} (y_1) = -1 \quad \text{in} \quad (a_i, a_{i+1}) \; . \end{cases} \tag{2.21}$$

Notice that (A2.6) or (A2.7) imply that N is even.

In the interior of each interval $\gamma_{i,j}$ the function $y_j(\lambda)$ is differentiable and

$$\frac{du}{dy} = \left(\frac{dy_j}{d\lambda}\right)^{-1} \quad \text{in} \quad \gamma_{i,j} \quad . \tag{2.22}$$

Because of (2.21) the function $y^*(\lambda)$ is differentiable in D_i^* and, using (2.20), one can show

$$\frac{du^*}{dy} = \left(\frac{dy^*}{d\lambda}\right)^{-1} \quad \text{in} \quad D_i^* \tag{2.23}$$

and

$$\frac{dy^*}{d\lambda} = -\frac{1}{2} \sum_{j=1}^{N} \left|\frac{dy_j}{d\lambda}\right| \quad \text{in} \quad (a_i, a_{i+1}) \quad . \tag{2.24}$$

In order to prove (G2b) we have to show

$$\int_{D_i} F(u) \; G\left(\left|\frac{du}{dy}\right|\right) dy \; \geq \; 2 \int_{D_i^* \cap \mathbb{R}_o^+} F(u^*) \; G\left(\left|\frac{du^*}{dx}\right|\right) dy$$

$$\text{for} \quad i = 1, \ldots, M-1 \quad . \tag{2.25}$$

Let us fix i and introduce λ as variable of integration. This is possible since $\frac{du}{dy} \left(\frac{du^*}{dx}\right)$ does not vanish in D_i (D_i^*) . Hence it remains to show

$$\sum_{j=1}^{N} \int_{\gamma_{i,j}} F(u) \; G\left(\left|\frac{du}{dy}\right|\right) dy = \sum_{j=1}^{N} \int_{a_i}^{a_{i+1}} F(\lambda) \; G\left(\left|\frac{dy_j}{d\lambda}\right|^{-1}\right) \left|\frac{dy_j}{d\lambda}\right| d\lambda \geq$$

$$\geq \int_{a_i}^{a_{i+1}} F(\lambda) \; G\left(2\left[\sum_{j=1}^{N} \left|\frac{dy_j}{d\lambda}\right|\right]^{-1}\right) \left[\sum_{j=1}^{N} \left|\frac{dy_j}{d\lambda}\right|\right] d\lambda \quad , \tag{2.26}$$

which is true, provided we can prove

$$\sum_{j=1}^{N} \alpha_j \, G\left(\left|\frac{dy_j}{d\lambda}\right|^{-1}\right) \geq G\left(2\left[\sum_{j=1}^{N} \left|\frac{dy_j}{d\lambda}\right|\right]^{-1}\right) , \tag{2.27}$$

where

$$\alpha_j := \left|\frac{dy_j}{d\lambda}\right|\left[\sum_{j=1}^{N} \left|\frac{dy_j}{d\lambda}\right|\right]^{-1} \quad \text{and} \quad \sum_{j=1}^{N} \alpha_j = 1 .$$

But by the convexity of G we have

$$\sum_{j=1}^{N} \alpha_j \, G\left(\left|\frac{dy_j}{d\lambda}\right|^{-1}\right) \geq G\left(\sum_{j=1}^{N} \alpha_j \left|\frac{dy_j}{d\lambda}\right|^{-1}\right)$$

$$= G\left(N\left[\sum_{j=1}^{N} \left|\frac{dy_j}{d\lambda}\right|\right]^{-1}\right) , \tag{2.28}$$

and for strictly convex G equality holds only if $\left|\frac{dy_j}{d\lambda}\right| = \left|\frac{dy_k}{d\lambda}\right|$ for $1 \leq j , k \leq N$. Now we recall that $N \geq 2$. This and the monotonicity of G imply

$$G\left(N\left[\sum_{j=1}^{N} \left|\frac{dy_j}{d\lambda}\right|\right]^{-1}\right) \geq G\left(2\left[\sum_{j=1}^{N} \left|\frac{dy_j}{d\lambda}\right|\right]^{-1}\right) , \tag{2.29}$$

and for strictly monotone G equality holds only if $N = 2$. Then (2.29) and (2.28) prove (2.27) and (G2b).

An inspection of the proof reveals that the equality in (G2b) implies that for all but a finite number of levels λ there are exactly two points y_1 and y_2 in Ω such that $u(y_1) = \lambda = u(y_2)$. Moreover $\frac{du}{dy}(y_1) = -\frac{du}{dy}(y_2) = 0$. But this implies ii).

Since nice functions are dense in $W^{1,p}(\Omega)$, one can extend part i) of Theorem 2.9 to functions in $W^{1,p}(\Omega)$. To this end it is necessary to extend the definition of u^* from Lipschitz continuous to $W^{1,p}$-functions.

Corollary 2.10

Let u be a function in $W^{1,p}(\Omega)$, $1 < p < \infty$, and suppose that (A2.5b) and at least one of (A2.6) or (A2.7) holds. Then (G1b) holds.

$$\int_{\Omega} \left|\frac{du}{dy}\right|^p dx \geq \int_{\Omega^*} \left|\frac{du^*}{dy}\right|^p dx \quad . \tag{G1b}$$

Without loss of generality we may in fact assume (A2.6). Otherwise we can reduce (A2.7) to (A2.6) as in Remark 2.24. For the proof we approximate u by a sequence $\{u_n\}_{n\in\mathbb{N}}$ of nice functions in $W_o^{1,p}(\Omega)$ such that each u_n satisfies (A2.6).

The rearranged sequence $\{u_n^*\}_{n\in\mathbb{N}}$ is bounded in $W_o^{1,p}(\Omega)$ and has a subsequence $\{u_{n_k}\}_{k\in\mathbb{N}}$ which converges weakly in $W_o^{1,p}(\Omega)$ to a limit v^* . Because of property (M9) $u_{n_k}^* \to u$ strongly in $L^p(\Omega)$, so $v^* = u^*$ in $L^p(\Omega)$. It remains to show that $v^* = u^*$ in $W_o^{1,p}(\Omega)$. To this end we observe that for any test function $\varphi \in C^\infty(\Omega)$ the relations

$$\int_{\Omega^*} \nabla v^* \, \nabla\varphi \, dx \leftarrow \int_{\Omega^*} \nabla u_{n_k}^* \, \nabla\varphi \, dx = -\int_{\Omega^*} u_{n_k}^* \, \Delta\varphi \, dx \to -\int_{\Omega^*} u^* \, \Delta\varphi \, dx =$$

$$= \int_{\Omega^*} \nabla u^* \, \nabla\varphi \, dx \tag{2.30}$$

hold. Hence $\nabla u^* = \nabla v^*$. Notice that this proof makes use of the boundary condition (A2.6). It is a modification of a proof in [73].

Remark 2.28

Another proof of Corollary 2.10 would follow from the continuity of the mapping $u \to u^*$ as a mapping in $W^{1,p}(\Omega)$. For $\Omega = \mathbb{R}$ and nonnegative functions on $\mathbb{R}$ this continuity was established in [56].

What about the second statement of Theorem 2.9? It can either be extended to simple functions, as the following Example 2.3 shows, nor to nonnegative $C_o^\infty(\Omega)$ functions. Nevertheless a weaker statement than Theorem 2.9 ii) holds for simple or smooth functions which satisfy (A2.6).

Under the assumptions of Theorem 2.9 ii) on G and F equality in (G2b) implies that u is quasiconcave. Recall that a continuous function $u : \overline{\Omega} \to \mathbb{R}$ on a convex set $\overline{\Omega}$ is called quasiconcave if $u\left(\frac{1}{2}x_1 + \frac{1}{2}x_2\right) - \min\{u(x_1), u(x_2)\} \geq 0$ for every $(x_1,x_2) \in \overline{\Omega} \times \overline{\Omega}$.

Example 2.3

Let u be defined on the interval $[-2,2]$ and

$$u(y) := \begin{cases} 2+y & \text{for} \quad -2 \le y \le -1 \,, \\ 1 & \text{for} \quad -1 \le y \le 0 \,, \\ 1+y & \text{for} \quad 0 \le y \le 1/2 \,, \\ 2-y & \text{for} \quad 1/2 \le y \le 2 \,. \end{cases}$$

Then equality holds in Corollary 2.10 but $u \neq u^*$.

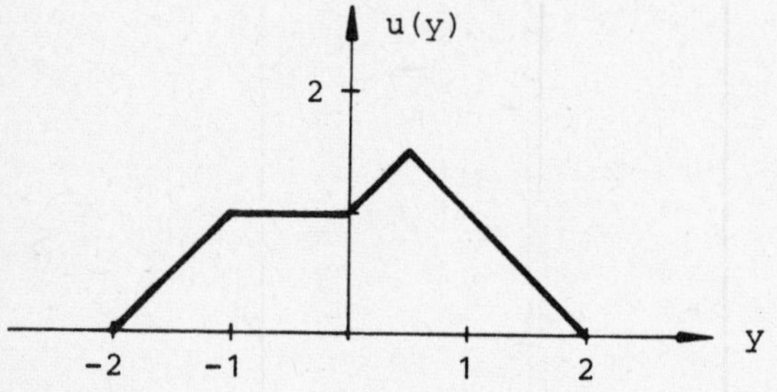

Figure 2.5

Remark 2.29

There is an extension of the concept of symmetric decreasing rearrangement in one dimension. Suppose $\Omega = (-\omega,\omega)$ and u is m-times periodic in Ω , $u(-\omega) = u(\omega)$. Then one can replace u by an m-times periodic symmetric function which is generated from the inverse of $\frac{1}{2m}\rho$ by periodic extension. Then analogues of Theorem 2.9 and Corollary 2.10 hold, and their proofs one has to observe that $N \geq 2m$. Results of this type and applications can be found in [31, 166].

II.5 Monotone decreasing rearrangement in direction y

We begin this paragraph with a conjecture of J. Rauch, which was the author's original motivation to study rearrangement. We hope that this conjecture will raise enough interest in the reader to endure another cycle of technical proofs.

Example 2.4

A conjecture of J. Rauch on hot spots or car acoustics.

Suppose $\Omega \subset \mathbb{R}^n$ is a bounded domain with boundary $\partial\Omega$ of class $C^{0,1}$. Consider the second eigenfunction of the Laplace operator under homogeneous Neumann boundary conditions. The first eigenvalue ν_1 is zero and the first eigenfunction u_1 is constant. The second eigenfunction u_2 satisfies

$$\Delta u_2 + \nu_2 u_2 = 0 \quad \text{in } \Omega$$

$$\frac{\partial u_2}{\partial n} = 0 \quad \text{on } \partial\Omega \quad .$$

It can be characterized as a solution to the following variational problem which is attributed to Lord Rayleigh [19, p. 101].

Minimize

$$R_1(v) = \frac{\int_\Omega |\nabla v|^2 \, dx}{\int_\Omega v^2 \, dx} \; ; \text{ over } \quad 0 \neq v \in W^{1,2}(\Omega) \; , \; \int_\Omega v \, dx = 0 \quad . \tag{2.31a}$$

In terms of physics u_2 represents an acoustical standing wave.

Conjecture: The function u_2 attains its maximum and minimum on the boundary $\partial\Omega$ (only).

Remark 2.30

J. Rauch was led to this conjecture in the following way. Consider the linear heat equation $u_t - \Delta u = 0$ in $\Omega \times \mathbb{R}_0^+$ under Neumann boundary conditions $\frac{\partial u}{\partial n} = 0$ on $\partial\Omega \times \mathbb{R}_0^+$ and with given Cauchy data $u(x,0) = u_0(x)$ in Ω. For each $t \in \mathbb{R}^+$ look at $X_t := \{z \in \overline{\Omega} \mid u(x,t)$ attains its (spatial) maximum at $z\}$. Loosely speaking the set X_t represents the hottest spots of the temperature distribution u at any given time t. According to J. Rauch it has been observed that the "hot spots move to the boundary" as $t \to \infty$. How could one prove this observation? If we separate the x and t variable, the solution $u(x,t)$ of this linear problem can be written as

$$u(x,t) = \sum_{j=1}^{\infty} \alpha_j e^{-\nu_j t} u_j(x) , \qquad (2.31b)$$

where ν_j is the j-th eigenvalue and u_j the j-th eigenfunction associated with the stationary problem. Since the boundary is thermally insulated, $u(x,t)$ will tend to the mean temperature $\int_\Omega u_o \, dx$. The first nonconstant eigenfunction in (2.31b), however, will "dominate" the "shape" of u for large t. Generically, i.e. almost always in physics, we can assume that $\alpha_2 = \int_\Omega u_o(x) \, u_2(x) \, dx$ does not vanish, so that one would expect the migration of the hot spot to $\partial\Omega$ as a consequence of the above conjecture. The preceding arguments are only heuristic and are meant to explain the term hot spot. Incidentally the conjecture is known to hold for special domains such as parallelepipeds, balls and annuli in $\mathbb{R}^n$, and there is numerical evidence for it if Ω is a plane polygonal nonconvex set [19, p. 122; 57, Vol. I, 5§5; 141, pp. 113, 127, 129; 154; 168, p. 301].

There may be a way to prove this conjecture using the inequality

$$\nu_2 < \lambda_1 . \qquad (2.32)$$

where λ_1 is the first eigenvalue of the associated Dirichlet problem.

Incidentally the relation (2.32) is based on two famous isoperimetric inequalities. Among all domains Ω of given n-dimensional Lebesgue measure λ_1 becomes minimal and ν_2 maximal for the ball.

An interesting consequence of (2.32) was observed by A. Pleijel: The "nodal set" $\{z \in \Omega | \; u_2(z) = 0\}$ cannot be a closed $(n-1)$ dimensional surface [19, p. 128; 148]. Unfortunately this proof does not extend to level surfaces $\{z \in \Omega | \; u_2(z) = \lambda\}$ for λ other than zero.

In this paragraph we shall prove the conjecture for cylindrical domains $\Omega = D \times (0,\omega)$ in $\mathbb{R}^n$ as Corollary 2.15. The idea of the proof comes from looking at the one dimensional case.Suppose $\Omega = (0,\omega)$ and we ignore the fact that u_2 has to be a cosine. Then one can prove the conjecture using (2.31a) and properties (C') and (G2a), (Lemma 2.6 ii)). If u would have a global maximum in Ω, then one could replace it by u^*. This argument can and will be generalized to cyclindrical domains, once we can establish property (G2d) (Theorem 2.8 ii)) for monotone decreasing rearrangement in direction y. We want to point out already that we have to discuss the equality sign in (G2), since the function u_2 need not to be unique. The nondegeneracy of ν_2 is known e.g. for cubes or balls.

Let us now prove some properties of the monotone decreasing rearrangement in direction y. We use the notation and definition from § II.1d). It is not obvious that D^* is closed if D is closed.

Lemma 2.11 Let $D \subset \mathbb{R}^n$ be bounded and closed. Then D^* is closed.

Proof: For $D = \emptyset$ the proof is trivial. If D is not empty, let $(x_k', y_k)_{k\in\mathbb{N}}$ be a sequence in D^* which converges to $(x',y) \in \mathbb{R}^n$ as $k \to \infty$. We have to show that $(x',y) \in D^*$ or equivalently, that $D(x')$ (cf. p. 11) is nonempty and that $|y| \le m_1(D(x'))$. If $D(x')$ were empty then there would be a neighborhood $V(x') \in \mathbb{R}^{n-1}$ such that for each $\tilde{x}' \in V(x')$ $D(\tilde{x}') = \emptyset$, since the complement of D is open in $\mathbb{R}^n$. This contradicts the fact that $(x_k',y_k) \in D$, because $D(x_k') \neq \emptyset$ for every $k \in \mathbb{N}$. Using the fact that χ_D is upper semicontinuous we can show

$$
\begin{aligned}
|y| &= \lim_{k\to\infty} |y_k| \le \lim m_1 (D(x_k')) \le \overline{\lim} \int_{\mathbb{R}} \chi_D(x_k',y)\, dy \le \\
&\le \int_{\mathbb{R}} \overline{\lim}\, \chi_D (x_k',y)\, dy \le \int_{\mathbb{R}} \chi_D (x',y)\, dy = \\
&= m_1 (D(x))
\end{aligned}
$$

as desired. Notice that we have used Fatou's lemma.

From now on let us assume in addition to (A2.1) that Ω is a cylindrical domain,

$$
\Omega = \Omega' \times (0,\omega), \text{ where } \Omega' \to \mathbb{R}^{n-1} \text{ is a bounded domain with boundary } \partial\Omega' \text{ of class } C^{0,1}. \tag{A2.9}
$$

This assumption is made until the end of Example 2.6. Then we can prove property (2L).

Lemma 2.12 Under the assumptions (A2.1) and (A2.9) if $u : \overline{\Omega} \to \mathbb{R}$ is Lipschitz continuous with Lipschitz constant L, then u is Lipschitz continuous with Lipschitz constant $2L$.

The proof of Lemma 2.12 is a modification of the one by M. Marcus [131]. Let $z_1, z_2 \in \overline{\Omega} = \overline{\Omega}^*$, $z_i = (x_i', y_i)$.

We intend to show $|u^*(u_1) - u^*(z_2)| \leq 2L\ |z_1 - z_2|$. Without loss of generality we may assume $u^*(z_1) < u^*(z_2)$, so that we have to show

$$u^*(z_1) \geq u^*(z_2) - 2L\ |z_1 - z_2| \quad . \tag{2.33}$$

To this end we define $E_2 := \{z = (x_2', y) \in \overline{\Omega}^* |\ u(z) \geq u^*(z_2)\}$, then $E_2^* := \{z \in \overline{\Omega}^* |\ u^*(z) \geq u^*(z_2)\}$ and $z_2 \in E_2^*$. Furthermore we define $E_1 := \{z = (x_1', y) \in \overline{\Omega}^* |\ d(z, E_2) \leq |z_1 - z_2|\}$.

Notice that here the proof differs from the one of Lemma 2.8 and from the one in [131]. By definition for every $z \in E_1$ there exists a $\tilde{z} \in E_2$ such that $u(\tilde{z}) \geq u^*(z_2)$ and $|\tilde{z} - z| \leq |z_1 - z_2|$. Hence $u(z) - u^*(z_2) = u(z) - u(\tilde{z}) + u(\tilde{z}) - u^*(z_2) \geq u(z) - u(\tilde{z}) \geq -L\ |z - \tilde{z}| \geq$ $\geq -L\ |z_1 - z_2|$.

This implies that

$$E_1 \subset \{z = (x_1', y) \in \overline{\Omega}^* |\ u(z) \geq u^*(z_2) - L\ |z_1 - z_2|\} .$$

Again we recall that rearrangements are order preserving, so that

$$E_1 \subset \{z = (x_1', y) \in \overline{\Omega}^* |\ u^*(z) \geq u^*(z_2) - L\ |z_1 - z_2|\} .$$

This time the "length" $m_1(E_1^*)$ merely exceeds the "length" $m_1(E_2^*)$, so that $d(z_1, E_1^*) \leq d(z_1, E_2^*) \leq |z_1 - z_2|$. Hence there exists a $\approx\!\!\!\!z \in E_1^*$ such that $|z_1 - \approx\!\!\!\!z| \leq |z_1 - z_2|$. Now we recall that $u^*(x_1', y)$ is Lipschitz continuous in direction y due to Lemma 2.3.

$u^*(z_1) - u^*(\approx\!\!\!\!z) \geq -L\ |z_1 - z_2|$, but $u^*(\approx\!\!\!\!z) \geq u^*(z_2) - L\ |z_1 - z_2|$ since $\approx\!\!\!\!z \in E_1^*$. This implies (2.33).

Remark 2.31

The factor 2 has purely technical reasons. They can be dealt with, for instance if $u(x,y) \geq u(x,\omega)$ for every $x \in \Omega'$, $y \in [0,\omega]$, because then one can extend u to $\overline{\Omega}' \times \mathbb{R}_0^+$ and follow the arguments of Lemma 2.8. Besides, Remark 2.24 applies correspondingly, so u^* is in fact Lipschitz continuous with Lipschitz constant L as a consequence of Corollary 2.14.

As for monotone decreasing rearrangement a) we define simple, nice, and smooth functions.

In this paragraph we call a function $u : \overline{\Omega} \to \mathbb{R}$ <u>simple</u> if and only if Ω satisfies (A2.5d), if $u \in C(\overline{\Omega})$ and if u is piecewise linear in the sense of affine.

We call a function $u : \overline{\Omega} \to \mathbb{R}$ <u>nice</u> if and only if it is simple and if $\frac{\partial u}{\partial y} \neq 0$ a.e. in Ω.

In this paragraph we call a function $u : \overline{\Omega} \to \mathbb{R}$ <u>smooth</u> if and only if $u \in C^1(\overline{\Omega})$, if Ω satisfies (A2.5d) and if u has the following properties: There exists a closed set $N \subset \overline{\Omega}'$ with $(n-1)$ dimensional Lebesgue measure zero such that

i) for every $x' \in \Omega' \setminus N$ and every $c \in \left(\min\limits_{\Omega} u \;,\; \max\limits_{\Omega} u\right)$, the set of points $\{(x',y) \in \Omega \mid u(x',y) = c\}$ is finite,

ii) for every $x' \in \Omega' \setminus N$ the set of points $\left\{(x',y) \in \Omega \,\middle|\, \frac{\partial u}{\partial y}(x',y) = 0\right\}$ is finite.

<u>Remark 2.32</u>

If we assume (A2.9) and Ω' polyhedral, nice functions are dense in $W^{1,p}(\Omega)$, $1 \leq p < \infty$. To see this we can approximate a function $u \in W^{1,p}(\Omega)$ by a function $\varphi \in C^2(\overline{\Omega})$ [3, p. 54]. Then we can partition Ω into finitely many nondegenerate n-simplices D_j, $j = 1, \ldots, N$, where the sets D_j are the convex hull of $n+1$ points $P_{j,k}$, $k = 1, \ldots, n+1$. For $n = 2$ they are triangles, for $n = 3$ tetrahedrons etc., and they can be chosen in such a way that the ratio h_j/r_j is bounded below by a positive constant α. Here $h_j = \operatorname{diam} D_j$ and r_j is the radius of the largest ball contained in D_j. Furthermore by changing N we can control $h = \max \{h_j\}_{j=1,\ldots,N}$.

Now we replace φ by its Lagrange interpolation φ_h, i.e. by a simple function which coincides with φ in all points $P_{j,k}$. Then it is well known [54] that $\|\varphi-\varphi_h\|_{W^{1,\infty}(\Omega)} \leq C\,h$, where C depends on α and φ but not on N. Therefore by choosing h small enough φ can be approximated in $W^{1,\infty}(\Omega)$ and thus in $W^{1,p}(\Omega)$ by simple functions. If a simple function w is not nice, it can be approximated by $w_\varepsilon(x,y) = w(x,y) + v_\varepsilon(y)$, where v_ε is defined in Remark 2.25.

Remark 2.33

In contrast to the one dimensional monotone decreasing rearrangement a), the proof of the following theorem works only for nice, but not for simple functions. The reason is the fact that the vanishing of $\frac{\partial u}{\partial y}$ on an open set of positive measure does not imply that ∇u is equal to zero on such a set.

Theorem 2.13

Let Ω satisfy (A2.9) and let the function u be smooth or nice. Let $F : \Omega' \times \text{range } u \to \mathbb{R}_o^+$ be nonnegative and continuous and let $G : \mathbb{R}_o^+ \to \mathbb{R}$ be monotone nondecreasing and convex.

i) Then the following inequality holds

$$\int_{\Omega} F(x',u)\, G(|\nabla u|)\, dx'\, dy \;\geq\; \int_{\Omega^*} F(x',u^*)\, G(|\nabla u^*|)\, dx'\, dy \quad . \tag{G2d}$$

ii) If moreover F is positive and G monotone increasing then equality holds in (G2d) only if u is monotone in y .

Remark 2.34

It is possible to extend the proof of this theorem to the case that G depends on a modified gradient

$$\sum_{k=1}^{n-1} X_k(x') \left|\frac{\partial u}{\partial x_k}(x',y)\right|^2 + X_o(x') \left|\frac{\partial u}{\partial y}(x',y)\right|^2 ,$$

where the coefficients X_k are positive and continuous.

For pedagogical reasons we omit these technical details here. They will be worked out in the case of Steiner symmetrization, though.

Under additional assumptions on u , i.e. $u(x,0) = 1$, $u(x,\omega) = 0$, $0 \leq u(x,y) \leq 1$, Theorem 2.13 i) was essentially derived by M. Marcus.

Open problem

How far can one weaken the smoothness assumptions on u and still prove Theorem 2.13 ii) ?

Corollary 2.14

Let Ω satisfy (A2.9), let Ω' be polyhedral and let $u \in W^{1,p}(\Omega)$, $1 < p < \infty$. Then $u^* \in W^{1,p}(\Omega)$ and we have

$$\int_{\Omega} |\nabla u|^p \, dx' \, dy \geq \int_{\Omega^*} |\nabla u^*|^p \, dx' \, dy \quad . \tag{G1d}$$

The proof follows from the approximation arguments in Remark 2.32. A function u can be approximated in $W^{1,p}(\Omega)$ by a sequence of function $\{u_n\}_{n\in\mathbb{N}}$. The rearranged sequence $\{u_n^*\}$ converges strongly in L^p to u^* and is bounded in $W^{1,p}(\Omega)$. After passing to a subsequence $\{u_{n_k}\}_{k\in\mathbb{N}}$ it converges weakly to a limit v^*. It remains to show that $v^* = u^*$ in $W^{1,p}(\Omega)$, then (G1d) follows from the weak lower semicontinuity of $\int_{\Omega} |\nabla u|^p \, dx' \, dy$ in $W^{1,p}(\Omega)$. But for any testfunction $\varphi \in C^{\infty}(\Omega)$ we have

$$\int_{\Omega} \nabla u \, \nabla\varphi \, dx' \, dy \leftarrow \int_{\Omega} \nabla u_{n_k}^* \, \nabla\varphi \, dx' \, dy = -\int_{\Omega} u_{n_k}^* \, \Delta\varphi \, dx' \, dy + \int_{\partial\Omega} u_{n_k}^* \, \frac{\partial\varphi}{\partial n} \, d\sigma \rightarrow$$

$$\rightarrow -\int_{\Omega} u^* \, \Delta\varphi \, dx' \, dy + \int_{\partial\Omega} u^* \, \frac{\partial\varphi}{\partial n} \, d\sigma = \int_{\Omega} \nabla u^* \, \nabla\varphi \, dx' \, dy \quad .$$

Notice that here we have used the continuity and linearity of the trace operator [3, p. 216], since it implies the weak convergence of $u_{n_k}^*$ to u^* in $L^p(\partial\Omega)$.

Proof of Theorem 2.13

The function u^* is Lipschitz continuous and hence the right hand side of (G2d) is well defined. We have to show that for almost every $x' \in \Omega'$ the following inequality holds

$$\int_0^{\omega} F(x',u) \, G(|\nabla u|) \, dy \geq \int_0^{\omega} F(x',u^*) \, G(|\nabla u^*|) \, dy \quad . \tag{2.34}$$

If u is nice then it is differentiable except on a finite set of polyhedral surfaces. Hence for almost every $x' \in \Omega'$ the function $u(x',y)$ is piecewise linear in y and except for finitely many points $(x',y_k)_{k=1,\dots,M(x')}$ differentiable with respect to x' and y.

If u is smooth, then by definition for almost every $x' \in \Omega'$ there are at most finitely many, say $M(x')$ points y_k, in which $\frac{\partial u}{\partial y}(x',y_k) = 0$.

For such a fixed x' let $a_1 \leq \dots \leq a_M$ be the values assumed by u at these M points. Define $D_i := \{y \in (0,\omega) \mid a_i < u(x',y) < a_{i+1}\}$ and $D_i^* := \{y \in (0,\omega) \mid a_i < u^*(x',y) < a_{i+1}\}$ for $i = 1, \dots, M-1$.

It suffices to integrate over these sets in (2.34). Now fix i and decompose D_i into a finite number of intervals $\{\gamma_{i,j}\}_{j=1,\dots,N(i,x')}$ in each of whose interior u is differentiable (with respect to all variables) and $\frac{\partial u}{\partial y} \neq 0$. For each $\lambda \in (a_i, a_{i+1})$ denote by $y_j(\lambda,x')$ the unique value of y in $\gamma_{i,j}$ for which $u(x',y_j) = \lambda$. The function $u^*(x',y)$ is monotone decreasing in y. Denote by $y^*(\lambda,x')$ the unique nonnegative value of y in D_i^* for which $u^*(x',y^*(\lambda,x')) = \lambda$. As in the proof of Lemma 2.6 we assume that the intervals $\{\gamma_{i,j}\}_{j=1,\dots,N}$ are ordered by their distance from the origin, so that

$$\operatorname{sign} \frac{\partial u}{\partial y}(x',y_j) = (-1)^{j+1} \operatorname{sign} \frac{\partial u}{\partial y}(x',y_1) \quad \text{in } \gamma_{i,j} \tag{2.35}$$

and

$$y^*(\lambda,x') = \begin{cases} \sum_{j=1}^{N} (-1)^j \, y_j(\lambda,x') & \text{if } \frac{\partial y_1}{\partial \lambda} > 0 \text{ and } \frac{\partial y_N}{\partial \lambda} < 0 , \\ \sum_{j=1}^{N} (-1)^{j+1} \, y_j(\lambda,x') & \text{if } \frac{\partial y_1}{\partial \lambda} < 0 \text{ and } \frac{\partial y_N}{\partial \lambda} < 0 , \\ \sum_{j=1}^{N} (-1)^j \, y_j(\lambda,x') + \omega & \text{if } \frac{\partial y_1}{\partial \lambda} > 0 \text{ and } \frac{\partial y_N}{\partial \lambda} > 0 , \\ \sum_{j=1}^{N} (-1)^{j+1} \, y_j(\lambda,x') + \omega & \text{if } \frac{\partial y_1}{\partial \lambda} < 0 \text{ and } \frac{\partial y_N}{\partial \lambda} > 0 . \end{cases} \tag{2.36}$$

We intend to replace the variable y by λ as independent variable in (2.34). Differentiation of $u(x',y_j(x',\lambda))$ with respect to λ and x_k yields

$$\frac{\partial u}{\partial y} = \left(\frac{\partial y_i}{\partial \lambda}\right)^{-1} \quad \text{in } \gamma_{i,j} , \tag{2.37}$$

$$\frac{\partial u}{\partial x_k} = - \frac{\partial y_j}{\partial x_k} \left(\frac{\partial y_j}{\partial \lambda}\right)^{-1} \quad \text{for } k = 1, \dots, n-1 \text{ in } \gamma_{i,j} , \tag{2.38}$$

and

$$|\nabla u| = \left|\frac{\partial y_j}{\partial \lambda}\right|^{-1} \left\{1 + \sum_{k=1}^{n-1} \left|\frac{\partial y_j}{\partial x_k}\right|^2\right\}^{1/2} \quad \text{in} \quad \Upsilon_{i,j} \ . \tag{2.39}$$

By the same arguments we obtain

$$\frac{\partial u^*}{\partial y} = \left(\frac{\partial y^*}{\partial \lambda}\right)^{-1} \qquad \text{in} \quad D_i^* \tag{2.40}$$

$$\frac{\partial u^*}{\partial x_k} = -\left(\frac{\partial y^*}{\partial x_k}\right)\left(\frac{\partial y^*}{\partial \lambda}\right)^{-1} \qquad \text{in} \quad D_i^* \quad \text{for} \quad k = 1, \ldots, n-1 \ , \tag{2.41}$$

and

$$|\nabla u^*| = \left|\frac{\partial y^*}{\partial \lambda}\right|^{-1} \left\{1 + \sum_{k=1}^{n-1} \left|\sum_{j=1}^{N} (-1)^j \frac{\partial y_j}{\partial x_k}\right|^2\right\}^{1/2} \quad \text{in} \quad D_i^* \ . \tag{2.42}$$

In the equation (2.36) we had to distinguish four different cases. In each case we calculate

$$\left|\frac{\partial y^*}{\partial \lambda}\right| = \sum_{j=1}^{N} \left|\frac{\partial y_j}{\partial \lambda}\right| \tag{2.43}$$

and

$$\left|\frac{\partial y^*}{\partial x_k}\right| = \left|\sum_{j=1}^{N} (-1)^j \frac{\partial y_j}{\partial x_k}\right| \ . \tag{2.44}$$

We have to prove

$$\sum_{j=1}^{N} \int_{\Upsilon_{i,j}} F(x',u)\, G(|\nabla u|)\, dy \geq \int_{o}^{\omega} F(x',u^*)\, G(|\nabla u^*|)\, dy \tag{2.45}$$

for every $i = 1, \ldots, M-1$. Let us fix i and introduce λ as a variable of integration. It remains to show

$$\sum_{j=1}^{N} \left|\frac{\partial y_j}{\partial \lambda}\right| G\left(\left|\frac{\partial y_j}{\partial \lambda}\right|^{-1} \left\{1 + \sum_{k=1}^{n-1} \left|\frac{\partial y_j}{\partial x_k}\right|^2\right\}^{1/2}\right) \geq$$

$$\geq \left(\sum_{j=1}^{N} \left|\frac{\partial y_j}{\partial \lambda}\right|\right) G\left(\left\{\sum_{j=1}^{N} \left|\frac{\partial y_j}{\partial \lambda}\right|\right\}^{-1} \left\{1 + \sum_{k=1}^{n-1} \left|\sum_{j=1}^{N} (-1)^j \frac{\partial y_j}{\partial x_k}\right|^2\right\}^{1/2}\right) \ . \tag{2.46}$$

We set $\alpha_j := \left|\frac{\partial y_j}{\partial\lambda}\right| \left(\sum_{j=1}^{N} \left|\frac{\partial y_j}{\partial\lambda}\right|\right)^{-1}$ and observe $\sum_{j=1}^{N} \alpha_j = 1$.

Since G is assumed to be convex and nondecreasing, a proof of (2.46) reduces to showing

$$\sum_{j=1}^{N} \left\{1 + \sum_{k=1}^{n-1} \left|\frac{\partial y_j}{\partial x_k}\right|^2\right\}^{1/2} \geq \left\{1 + \sum_{k=1}^{n-1} \left|\sum_{j=1}^{N} (-1)^j \frac{\partial y_j}{\partial x_k}\right|^2\right\}^{1/2} . \quad (2.47)$$

We interpret the left hand side of (2.47) as the sum over the L^2-norms of vectors $\vec{a}^j = (a_o^j, a_1^j, \ldots, a_{n-1}^j)$, where $a_o^j = 1$, $a_k^j = \left|\frac{\partial y_j}{\partial x_k}\right|$ for $k = 1, \ldots, n-1$. According to Minkowski's inequality this sum dominates the length of $\sum_{j=1}^{N} \vec{a}_j$. Hence we can verify (2.47)

$$\sum_{j=1}^{N} \|\vec{a}_j\| \geq \left\|\sum_{j=1}^{N} \vec{a}_j\right\| = \left\{\left(\sum_{j=1}^{N} 1\right)^2 + \sum_{k=1}^{n-1} \left(\sum_{j=1}^{N} \left|\frac{\partial y_j}{\partial x_k}\right|\right)^2\right\}^{1/2} \geq$$

$$\geq \left\{N^2 + \sum_{k=1}^{n-1} \left|\sum_{j=1}^{N} (-1)^j \frac{\partial y_j}{\partial x_k}\right|^2\right\}^{1/2} >$$

$$\geq \left\{1 + \sum_{k=1}^{n-1} \left|\sum_{j=1}^{N} (-1)^j \frac{\partial y_j}{\partial x_k}\right|^2\right\}^{1/2} .$$

An inspection of the proof yields that under the strengthened assumptions of F and G equality holds only if $N(i,x') = 1$ for $i = 1, \ldots, M-1$ and almost every $x' \in \Omega'$. This means that u is monotone in y .

Let us now return to Example 2.4.

Corollary 2.15

<u>If Ω satisfies (A2.9) and if u_2 is a solution of (2.31a) then u_2 attains its maximum and minimum on $\partial\Omega$.</u>

For the proof we observe that we can replace u_2 by u_2^* . u_2^* satisfies the orthogonality condition $\int u_2^* \, dx = \int u_2 \, dx = 0$ and $\int u_2^2 \, dx = \int (u_2^*)^2 \, dx$. If u_2 is smooth, Theorem 2.13 ii) implies that $\frac{\partial u_2}{\partial y}$ does not change sign, so that u_2 attains its maximum and minimum on $\{0\} \times \Omega'$ or $\{\omega\} \times \Omega'$. If $\frac{\partial u_2{}^2}{\partial y} \equiv 0$, then there is nothing to prove. It remains to show that either u_2 is smooth or u_2 is independent of y. Suppose that u_2 does depend on y. We observe that the function $\frac{\partial u_2}{\partial y}$ satisfies a homogeneous Dirichlet condition on $\{0\} \times \Omega'$ and $\{\omega\} \times \Omega'$ and has an analytic extension across these parts of $\partial\Omega$. This follows from standard regularity theory [139]. Smooth functions have to satisfy two properties i) and ii). Let us denote the set of those points $x \times \Omega'$ in which i) fails by N_1. I intend to show that either $N_1 = \Omega'$ in which $\frac{\partial u_2}{\partial y} \equiv 0$ in Ω or that N_1 is a closed set of $(n-1)$ dimensional measure zero. $N_1 \times (0,\omega)$ is certainly contained in $N_2 := \left\{(x,y) \in \Omega \;\middle|\; \frac{\partial u_2}{\partial y}(x',y) = 0\right\}$. N_2 consists of

$$\partial\left\{(x,y) \in \Omega \;\middle|\; \frac{\partial u_2}{\partial y}(x,y) \geq 0\right\} = \left\{z \in \Omega \;\middle|\; \frac{\partial u_2}{\partial y}(z) = 0\right.$$

$$\left.\text{and } \frac{\partial u_2}{\partial y} > 0 \text{ somewhere in every nonempty open neighborhood of } z\right\}$$

and of $\partial\{(x,y) \in \Omega \;|\; \frac{\partial u_2}{\partial y}(x,y) \leq 0\}$. Any other points in N_2 would be interior points, but then $\frac{\partial u_2}{\partial y}$ would be identical zero in Ω. Hence the set of points in which $\frac{\partial u_2}{\partial y}$ vanishes consists of piecewise analytic hypersurfaces of dimension at most $(n-1)$. This can be found e.g. in [19, p. 53].

If $x' \in N_1$, then N_2 contains the line segment $\{(x',y),\ 0 \leq y \leq \omega\}$, so that N_1 consists of piecewise analytic hypersurfaces of dimension at most $(n-2)$. Therefore, N_1 is closed and has $(n-1)$ dimensional measure zero as desired.

Remark 2.35

The set N_2 in the previous proof also consists of $N_{21} := \{(x,y) \in N_2 \;|\nabla u(x,y) \neq 0\}$ and $N_{22} := \{(x,y) \in N_2 \;|\nabla u(x,y) = 0\}$. N_{21} contains piecewise analytic hypersurfaces because of the implicit function theorem. Recently it was shown [44] that N_{22} contains a countable number of analytic $(n-2)$ dimensional manifolds.

Remark 2.36

Let us explain why we cannot prove Corollary 2.13 for convex domains.

If u_2 attains its maximum in z_M and its minimum in z_m one might expect u_2 to be monotone nonincreasing in the direction $z_m - z_M$, and without loss of generality this can be the direction y. Then one could suitably alter the definition of u^* in such a way that u^* is defined on Ω and monotone nonincreasing in direction y. It is possible to prove local Lipschitz continuity of u^* and to start as in the proof of Theorem 2.13, but the principal difficulty is the following: In the representation (2.36) ω would depend on x, so that (2.42) would contain derivatives of ω. These derivatives would show up on the right hand side of (2.46), but not on the left. Therefore there is little hope that this proof can be extended to convex domains. For nonconvex domains J. Hersch believes to have a counter-example to J. Rauchs conjecture [90].

Open problem

Prove J. Rauchs conjecture for convex domains.

Remark 2.37

The fact that $\Omega = \Omega' \times (0,\omega)$ is cyclindrical and problem (2.31a) is linear calls for a separation of variables. Let ν_2 and u_2 be the second eigenvalue and eigenfunction associated with Ω, ν_2' and u_2' the eigenvalues and eigenfunctions belonging to Ω' and $(\pi/\omega)^2$ and $\cos(\frac{\pi}{\omega} y)$ the eigenvalue and eigenfunction for the interval $(0,\omega)$. Then $v(x,y) = u_2'(x)$ and $w(x,y) = \cos(\frac{\pi}{\omega} y)$ are admissible functions for the Rayleigh quotient and we can draw some conclusions from this. First we note that $\nu_2 = \min\left\{\nu_2', \frac{\pi}{\omega}\right\}$. If $\nu_2' < \frac{\pi}{\omega}$, then $u_2 = c\, u_2'$, $c \in \mathbb{R}$, and u_2 is independent of y and attains its maximum at $y = 0$ and $y = \omega$. If $\nu_2' > \frac{\pi}{\omega}$, then $u_2 = c \cos(\frac{\pi}{\omega} y)$, $c \in \mathbb{R}$, and u_2 is independent of x and attains its maximum at $y = 0$ or $y = \omega$. Finally if $\nu_2' = (\frac{\pi}{\omega})^2$, then $u_2 = c_1 u_2' + c_2 \cos(\frac{\pi}{\omega} y)$, $c_i \in \mathbb{R}$ and u_2 still attains its maximum at $y = 0$ or $y = \omega$. This last conclusion relies on the separated structure of u_2 and was kindly pointed out to me by J.C.C. Nitsche. (If two functions v and w both attain their maximum on $\partial\Omega$, then a linear combination of v and w does not necessarily attain its maximum on $\partial\Omega$.)

Remark 2.38

If a domain Ω becomes cylindrical after a change of coordinates, and if the transformed gradient can be written as in Remark 2.34, then J. Rauchs conjecture is still true. This is the case e.g. for the "pacman domain" $\{(r,\varphi) \in \mathbb{R}^2 \mid 0 \leq r \leq 1,\ |\varphi| < 0.8\,\pi\}$ in $\mathbb{R}^2$, or for torodial sectors in $\mathbb{R}^3$ with arbitrary cross section Ω'.

Monotone decreasing rearrangement can be applied to some other problems as well, as can be seen from the following example.

Example 2.5

Front tracking for a free boundary problem. In [137] G.H. Meyer suggests a numerical algorithm to calculate free boundaries in nonlinear reaction-diffusion problems from chemical engineering. One of his problems could be stated as follows.

Minimize

$$J_2(v) \quad := \int_\Omega \left\{\frac{1}{2}\,|\nabla v|^2 + \alpha\, v - \alpha \ln\,(1+v) + \gamma\, x^2\, v\right\} dx\, dy \tag{2.48}$$

over the set

$$\mathbb{K}_2 \quad := \Big\{v \in W^{1,2}(\Omega)\,|\; v \geq 0 \;\text{ a.e. in }\; \Omega\;,\; v = g \;\text{ on }\; \Gamma_1\;,$$
$$v = 0 \;\text{ on }\; \Gamma_2 \;\text{ and }\; v \geq 0 \;\text{ on }\; \Gamma_3\Big\}\;.$$

Here α and γ are positive real numbers, Ω is the square $(-0.5,+0.5) \times (0,1)$ in $\mathbb{R}^2$, $\Gamma_1 := \{(x,0)\,|\; x \in (0.5,+0.5)\}$, $\Gamma_2 := \{(x,y)|\; x = +\,0.5$ and $y \in (0,1)\}$ and $\Gamma_3 = \{(x,y)\,|\; y = 1\;,$ $x \in (-0.5,+0.5)\}$. Finally $g(x,0) = \frac{1}{4} - x^2$.

Problems of this type were first studied in the pioneering paper of G. Fichera [71] and the existence, uniqueness and regularity of solutions can for instance be found in [18, 107]. It is known that problem (2.48) has a unique solution in $W^{2,2}(\Omega)$ which solves the free boundary problem

$$\Delta u \;=\; \frac{\alpha\, u}{1+u} + \gamma\, x^2 \quad \text{in} \quad \{(x,y) \in \Omega|\; u(x,y) > 0\}\;, \tag{2.49}$$

$$\left.\begin{aligned} u &= g \;\text{ on }\; \Gamma_1\;,\\ u &= 0 \;\text{ on }\; \Gamma_2\;,\\ u \geq 0\;,\; \frac{\partial u}{\partial n} &\geq 0 \;\text{ and }\; u \cdot \frac{\partial u}{\partial n} = 0 \;\text{ on }\; \Gamma_3\;,\\ u &= \frac{\partial u}{\partial n} = 0 \;\text{ on }\; \partial\,\{u > 0\} \cap \Omega \end{aligned}\right\} \tag{2.50}$$

in the classical sense.

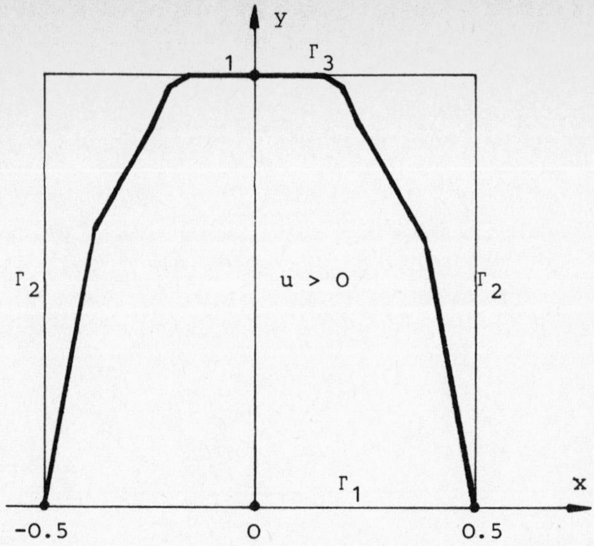

Figure 2.6

The shape of the free boundary $\partial\{u > 0\}$ is unknown.

G.H. Meyer assumes that it can be expressed as $y = S(x)$. This assumption is crucial for his calculations and, using monotone decreasing rearrangement in direction y , one can show that it is indeed satisfied. In fact one can approximate u in $W^{1,2}(\Omega)$ by a minimizing sequence $\{U_n\}_{n\in\mathbb{N}}$ of nice functions. The rearranged sequence $\{U_n^*\}_{n\in\mathbb{N}}$ has a weakly convergent subsequence in $W^{1,2}(\Omega)$ which converges pointwise to u^* almost everywhere. One can easily check that u^* is again in $\mathbb{K}_2$. To see that $u^* = g$ on Γ_1 one introduces $D(x,y) = u(x,0) - u(x,y)$ for $(x,y) \in \Omega \cap \{u > 0\}$. If D attains a negative minimum in $\Omega \cap \{u > 0\}$, then $\Delta D(x,y) \geq 0$, i.e. $u_{xx}(x,0) = g_{xx} = -2 \geq \Delta u(x,y) > 0$, a contradiction. D cannot attain a negative minimum on $\partial\{u > 0\} \cap \Omega$, nor on Γ_1 or Γ_2 . If D attains a negative minimum on Γ_3 , then $D_y(x,y)$ has to be negative, i.e. $u_y(x,y)$ is positive and using Signorini's boundary condition $u(x,y) = 0$, another contradiction. So u^* is in fact in $\mathbb{K}_2$ and by uniqueness u has to equal u^* and u is monotone nonincreasing in y . This implies that the free boundary $\partial\{u > 0\} \cap \Omega$ cannot be intersected twice by a line segment $\{x = \text{const}, y \in (0,1)\}$. It is however conceivable that the free boundary contains a vertical line segment

$\{x = \text{const}, y \in (a,b)\}$, const $\in (-0.5,+0.5)$.

Corollary 2.16

The free boundary in problem (2.48) can be expressed as a multivalued function $y \in S(x)$ where $S(x)$ is simply connected for each $x \in (-0.5,+0.5)$. If the free boundary is analytic, it can be expressed as $y = S(x)$.

The first statement in this corollary was derived above. The second statement holds if the free boundary does not contain a vertical line segment $\{x = c, y \in (a,b)\}$, $c \in (0,1)$, $a < b$. If it does, then u could be analytically extended from its support across the free boundary , and thus $u(c,0)$ would have to be zero, a contradiction.

The calculations of Meyer suggest that u is symmetric in x . This is a consequence of its uniqueness. Using Steiner symmetrization in direction x one can derive additional information on the shape of the free boundary, but this can also be done by means of monotone decreasing rearrangement.

To this end we restrict u to those $(x,y) \in \Omega$ with positive x component, replace it by its monotone decreasing rearrangement in direction x and reflect the rearranged function across $\{x = 0\}$. This way we obtain a function $\tilde{u}$ which is symmetrically decreasing in x . Since u was symmetric in x , u and $\tilde{u}$ are equimeasurable and $J(\tilde{u}) \leq J(u)$. Therefore $u = \tilde{u}$. Hence we have shown

Corollary 2.17

The free boundary in problem (2.48) is starshaped with respect to the origin.

Example 2.6 The dam problem

Suppose an earthen dam of rectangular crossection separates two reservoirs of water at different levels $y = H$ and $y = h < H$ from each other. Since the dam is porous, there is water in the dam and the dam is partly wet. The study of the wet region leads after a tricky substitution ("Baiocchi's trick") to the following variational problem [17, 60, 75].

Minimize

$$J_3(v) \quad = \quad \int_\Omega \{|\nabla v|^2 - 2\, v\}\, dx\, dy \tag{2.51}$$

over

$$\mathbb{K}_3 \quad := \quad \left\{v \in W^{1,2}(\Omega) \mid v = g \quad \text{on} \quad \partial\Omega \;,\; v \geq 0 \quad \text{in} \quad \Omega\right\} \;.$$

Here

$$\Omega \quad := \quad (0,a) \times (0,H) \quad \text{and}$$

$$\begin{aligned} g(0,y) &= \frac{1}{2}\,(H-y)^2 && \text{on} \quad \Gamma_1 \;, \\ g(a,y) &= \frac{1}{2}\,(h-y)^2 && \text{on} \quad \Gamma_2 \;, \\ g(x,0) &= \frac{H^2}{2}\left(1 - \frac{x}{a}\right) + \frac{h^2}{2}\,\frac{x}{a} && \text{on} \quad \Gamma_o \;, \\ g &= 0 \qquad \text{elswhere} && \text{on} \quad \partial\Omega \;. \end{aligned} \tag{2.52}$$

The definition of Γ_i , $i = 0, 1, 2$ is obvious from Figure 2.7.

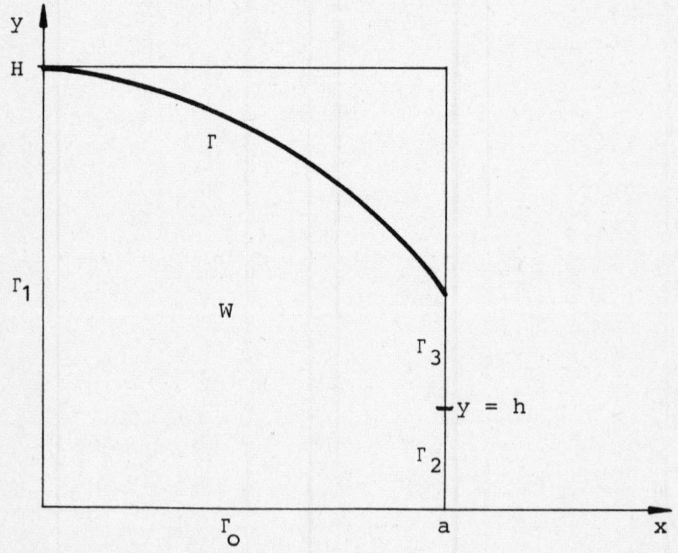

Figure 2.7

Let us denote a solution of problem (2.51) by $w(z)$.

The wet region of the dam is the set $\{z \in \Omega \mid w(z) > 0\}$ and its boundary in Ω is denoted by Γ . It is known that (2.51) has a unique solution in $W^{2,\infty}_{loc}(\Omega) \cap W^{2,p}(\Omega)$ for any $p < \infty$, and that the wet region is convex. The convexity proof contains several steps. One of the steps consists in showing that <u>the wet region is starshaped</u> with respect to the origin, and for this step we can give a <u>new proof</u>.

If u is a solution of problem (2.51) then we can repeat the arguments for the proof of Corollary 2.17 and conclude that u is monotone nonincreasing both in x and y , hence $x \; u_x + y \; u_y \leq 0$ in Ω . Therefore the wet region is starshaped with respect to zero.

<u>Open problem</u>

Let Ω be the unit square $(0,1) \times (0,1)$ in $\mathbb{R}^2$ and $f : \Omega \to \mathbb{R}$ a sufficiently smooth function with mean value zero. Consider the problems:

$$- \Delta u = f \text{ in } \Omega , \qquad - \Delta v = f^* \text{ in } \Omega ,$$

and

$$\frac{\partial u}{\partial n} = 0 \text{ on } \partial\Omega , \qquad \frac{\partial v}{\partial n} = 0 \text{ on } \partial\Omega ,$$

where f^* is the monotone decreasing rearrangement of f in direction y . In the authors opinion the oscillation of u over $\overline{\Omega}$ should be dominated by the oscillation of v . The oscillation of u over $\overline{\Omega}$ is defined as $\max\limits_{x \in \overline{\Omega}} u(x) - \min\limits_{x \in \overline{\Omega}} u(x)$.

II.6 Starshaped rearrangement

Starshaped rearrangement can be interpreted as a monotone decreasing rearrangement in direction r , where r denotes the radial components of n-dimensional spherical coordinates. Throughout this paragraph we assume:

$\overline{\Omega}$ is bounded and starshaped with respect to zero in short, $\overline{\Omega} = \overline{\Omega}^{*(p)}$ for every $p \geq 0$. (A2.5e)

$u : \mathbb{R}^n \to \mathbb{R}_o^+$ is a Lipschitz continuous function with compact support $D \subset \overline{\Omega}$. $\overline{\Omega}$ contains a nonempty open ε-neighborhood $U_\varepsilon(0)$ of the origin and u attains its maximum at each point of $U_\varepsilon(0)$. (A2.4)

For the definition of $\Omega^{*(p)}$ and $u^{*(p)}$ we refer to § II.1. Let us first show that u and $u^{*(o)}$ are equimeasurable.

<u>Lemma 2.18</u>

<u>If</u> D <u>contains</u> $U_\varepsilon(0)$ <u>and its compact, then</u> $m_n(D) = m_n(D^{*(o)})$.

For the proof we calculate $\int_D dx$ using polar coordinates. The Jacobian associated with the transformation from cartesian to polar coordinates can be written as

$$j(\theta)\, r^{n-1} := \frac{\partial(x_1,\ldots,x_n)}{\partial(r,\theta_1,\ldots,\theta_{n-1})} , \tag{2.53}$$

where $j(\theta)$ is a continuous function of θ . The Jacobian vanishes on a set of n-dimensional measure zero. This can be seen from Sard's theorem (e.g. [33, p. 52]) or from the fact that $j(\theta)$ vanishes only for particular angles and that r vanishes only in the origin. Therefore we can disregard the set where the Jacobian vanishes when we integrate [16, p. 78 f.]. Using the notation from p. 12 we have

$$m_n(D) = \int_D dx = \int_{\theta\in T} j(\theta)\, h(\theta)\, d\theta = \int_{\theta\in T} j(\theta)\, \frac{1}{n} R^n(\theta)\, d\theta =$$

$$= \int_{\theta\in T} j(\theta) \int_0^{R(\theta)} r^{n-1}\, dr\, d\theta = \int_{D^{*(o)}} dx = m_n(D^{*(o)}) .$$

For $n = 2$ Lemma 2.18 states that the area $m_2(D)$ of a set D is invariant under rearrangement. The general experience in the theory of rearrangement suggests that rearrangement decreases the perimeter of a set. This is in fact well known for Steiner - and Schwarz symmetrization [156]. The following counter-example, however, shows that we cannot expect this kind of result for starshaped rearrangement under the (equimeasurable) metric $g(r) = r^{n-1}$.

Theorem 2.19

There exists a compact domain $D \supset U_\varepsilon(0)$ in $\mathbb{R}^2$ such that the perimeter of D is shorter than the perimeter of $D^{*(o)}$.

The domain is a simplification of H. Grabmüller's "long nose" [86]. Let D be the union of an ε-neighborhood of zero, with $0 < \varepsilon << 1$ and of the annular sector $\{x = (r,\theta) \in \mathbb{R}^2 |\ 1 \leq r \leq 2\ ,\ 0 \leq |\theta| \leq \varphi\}$, where $\varphi \in (0,\pi)$ is determined below. The perimeter of D has length $|\partial D| = 2 + 6\varphi + 2\pi\varepsilon$; $D^{*(o)} = \{x = (r,\theta) \in \mathbb{R}^2 |\ 0 \leq r \leq \sqrt{3} + O(\varepsilon)\ ,\ 0 \leq |\theta| \leq \varphi\} \cup U_\varepsilon(0)$ and the perimeter of $D^{*(o)}$ has length $|\partial D^{*(o)}| = 2\sqrt{3} + \sqrt{3} \cdot 2\varphi + O(\varepsilon)$. A comparison of $|\partial D|$ and $|\partial D^{*(o)}|$ shows that $|\partial D^{*(o)}| > |\partial D|$ for sufficiently small values of φ .

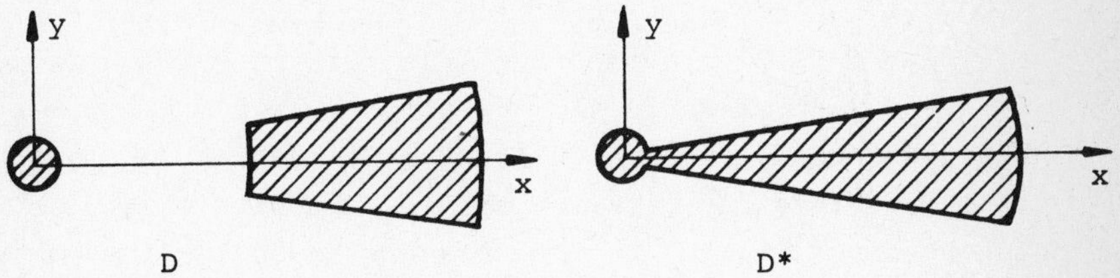

Figure 2.8

The construction of the set D is illustrated in Fig. 2.8. Theorem 2.19 can be used to show that inequality (G1) cannot hold for equimeasurable rearrangement.

Corollary 2.20

<u>There exists a domain</u> $\Omega \subset \mathbb{R}^2$ <u>and a function</u> $v \in W_o^{1,1}(\Omega)$ <u>which satisfy</u> (A2.5e) <u>and</u> (A2.4), <u>but for which</u> (G1) <u>fails, i.e.</u>

$$\int_\Omega |\nabla v^{(o)}|\, dx > \int_\Omega |\nabla v|\, dx \qquad \text{holds.} \tag{2.54}$$

Let Ω be a ball with radius 3 and center in the origin, and let D be the set constructed in Theorem 2.19. Now let $v : \Omega \to [0, \frac{\varepsilon}{2}]$ be defined by

$$v(x) := \begin{cases} 0 & \text{if } x \in \Omega \setminus D , \\ d(x,\partial D) & \text{if } x \in D \text{ and } d(x,\partial D) \leq \frac{\varepsilon}{2} , \\ \varepsilon/2 & \text{if } x \in D \text{ and } d(x,\partial D) > \frac{\varepsilon}{2} . \end{cases}$$

If ε and φ are sufficiently small we have the strict inequality $|\partial\Omega_c| < |\partial\Omega_c^{*(o)}|$ for $c \in (0,\varepsilon/2)$ according to Theorem 2.19.

But Federer's coarea formula [70] states that

$$\int_\Omega |\nabla w|\, dx = \int_{\mathbb{R}} P(t)\, dt , \text{ where } P(t) = \text{perimeter of } \{x \in \Omega |\ w(x) > t\} .$$

This implies (2.54).

As mentioned in § II.2 one can prove (G1) for nonequimeasurable starshaped rearrangement. Thus in view of potential applications in the calculus of variations the nonequimeasurable rearrangement appears to be natural. Therefore the question arises what kind of substitutes one has for properties (E) and (C). To this end we shall first prove a calculus lemma:

Lemma 2.21

<u>Let</u> $m \in \mathbb{N}_o$ <u>and</u> $0 < \varepsilon \leq r_1 < r_2 < \ldots \leq r_{2m+1}$, $r_j \in \mathbb{R}$, <u>let</u> $2 \leq n \in \mathbb{N}$ <u>and</u> $1 \leq p < \infty$, $p \neq n$. <u>Then the following inequalities hold:</u>

$$\left(r_1^{n-p} - r_2^{n-p} + \ldots + r_{2m+1}^{n-p}\right)^{1/(n-p)} \leq \left(r_1^n - r_2^n + \ldots + r_{2m+1}^n\right)^{1/n} , \tag{2.55}$$

$$\frac{r_1 \cdot r_3 \cdot r_5 \cdot \ldots \cdot r_{2m+1}}{r_2 \cdot r_4 \cdot \ldots \cdot r_{2m}} \leq \left(r_1^n + r_2^n + \ldots r_{2m+1}^n\right)^{1/n} . \tag{2.56}$$

First proof:

For the proof of (2.55) we have to distinguish the cases $p < n$ and $p > n$. The case of $p < n$ is documented in a paper by L. Payne and A. Weinstein [151] who attribute its proof to H. Weinberger. A more general inequality was also found by G. Szegö; we refer to [138, p. 112] for historical details. As it turns out their idea [151] of proof applies to both cases $p < n$, $p > n$ and to proving (2.56). If we introduce the notation $y_j := r_j^n$, $\alpha = \frac{n-p}{n}$, inequality (2.55) is equivalent to

$$F_1(y_1,\dots,y_{2m+1}) := (y_1^\alpha - y_2^\alpha + \dots - y_{2m+1}^\alpha)^{1/\alpha} - y_1 + y_2 - \dots - y_{2m+1} \leq 0 . \quad (2.57)$$

Inequality (2.57) can be established by introduction with respect to m. Obviously (2.57) holds for $m = 0$. Suppose it holds for $m - 1$, i.e.

$$F_1(y_1,\dots,y_{2m-1},y_{2m},y_{2m}) \leq 0 . \quad (2.58)$$

The fundamental theorem of calculus implies

$$F_1(y_1,\dots,y_{2m-1},y_{2m},y_{2m+1}) = F_1(y_1,\dots,y_{2m-1},y_{2m},y_{2m}) + \int_{y_{2m}}^{y_{2m+1}} \frac{\partial F_1}{\partial t}(y_1,\dots,y_{2m},t)\, dt \quad (2.59) ,$$

but a simple calculation shows that

$$\frac{\partial F_1}{\partial t}(y_1,\dots,t) = (y_1^\alpha - y_2^\alpha + \dots + t^\alpha)^{\frac{1-\alpha}{\alpha}} t^{\alpha-1} - 1 = \left[\left(\frac{y_1}{t}\right)^\alpha - \left(\frac{y_2}{t}\right)^\alpha + \dots + \left(\frac{t}{t}\right)^\alpha\right]^{\frac{1-\alpha}{\alpha}} - 1 . \quad (2.60)$$

Recall that $0 < y_1^\alpha < \dots \leq t^\alpha$.

If $p < n$, $\alpha > 0$ and the square bracket is bounded above by 1. But $\frac{1-\alpha}{\alpha} = \frac{p}{n-p} > 0$, so that $\frac{\partial F_1}{\partial t} \leq 0$ and hence (2.57) holds for m as well.

If $p > n$, $\alpha < 0$ and $\frac{1-\alpha}{\alpha} < 0$. This time $y_1^\alpha > y_2^\alpha > \dots \geq t^\alpha > 0$, so that the square bracket in (2.60) is bounded below by 1. Therefore, a negative power of this bracket is bounded above by 1 and (2.57) holds for $p > n$. Finally we have to prove (2.56), equivalently

$$F_2(r_1,r_2,\dots,r_{2m+1}) := \frac{r_1 \cdot r_3 \cdot \dots \cdot r_{2m+1}}{r_2 \cdot \dots \cdot r_{2m}} - \left(r_1^n - r_2^n + \dots + r_{2m+1}^n\right)^{1/n} \leq 0 . \qquad (2.61)$$

This will again be done by induction with respect to m. (2.61) holds for $m = 0$. Let it be true for $m - 1$, i.e.

$$F_2(r_1,r_2,\dots,r_{2m-1},r_{2m},r_{2m}) \leq 0 ,$$

$$F_2(r_1,\dots,r_{2m-1},r_{2m},r_{2m+1}) = F_2(r_1,\dots,r_{2m-1},r_{2m},r_{2m}) + \int_{r_{2m}}^{r_{2m+1}} \frac{\partial F_2}{\partial t}(r_1,\dots,r_{2m},t)\,dt$$

and calculate the derivative of F with respect to t:

$$\frac{\partial F_2}{\partial t}(r_1,\dots,t) = \frac{r_1 \cdot r_3 \cdot \dots \cdot r_{2m-1}}{r_2 \cdot r_4 \cdot \dots r_{2m}} - \left(r_1^n - r_2^n + \dots + t^n\right)^{\frac{1-n}{n}} t^{n-1} . \qquad (2.62)$$

Since $r_1 < r_2 < \dots \leq t$ the first term on the right hand side of (2.62) is less than 1 and

$$\left(r_1^n - r_2^n + \dots + t^n\right)^{\frac{1-n}{n}} t^{n-1} = \left[\left(\frac{r_1}{t}\right)^n - \left(\frac{r_2}{t}\right)^n + \dots + \left(\frac{t}{t}\right)^n\right]^{\frac{1-n}{n}} \geq 1 ,$$

because the square bracket is less than or equal to 1 and $\frac{1-n}{n}$ is negative. Therefore the derivative (2.62) is nonpositive and the proof is complete.

Second proof:

For $p < n$, see the paper of L. Payne and A. Weinstein, for $p > n$ apply [138, p. 112, Thm. 5], for $n = p$ apply l'Hospital's rule to (2.57).

Now we are in a position to prove a substitute for (E). We shall only state it for polyhedral domains, but it can be extended e.g. to domains with Lipschitz continuous boundary.

Lemma 2.22

If $D \subset \mathbb{R}^n$ is bounded and polyhedral and contains $U_\varepsilon(0)$, then for $p \geq 1$

$$D^{*(0)}(\Theta) \supset D^{*(p)}(\Theta) \text{ for almost every } \Theta \in T \text{, and } m_n(D) \geq m_n(D^{*(p)}) . \qquad (E')$$

We observe that for almost every $\Theta \in T$ the set D has the property:

The ray $\{\Theta = \text{const}, r \geq 0\}$ intersects ∂D in an odd number $2m(\Theta)+1$ of points $0 < \varepsilon \leq r_1(\Theta) < \ldots < r_{2m+1}(\Theta)$ (A2.3) with $m \in \mathbb{N}_0$.

We can neglect those Θ for which (A2.3) fails when we integrate over T. Furthermore we recall (2.3), (2.4) and (2.53) and see

$$m_n(D) = \int_{\Theta \in T} j(\Theta) \cdot \frac{1}{n} \cdot \left(r_1^n - r_2^n + \ldots + r_{2m+1}^n\right) d\Theta \quad ,$$

$$D^{*(0)}(\Theta) = \left\{(r,\Theta) \mid \varepsilon \leq r \leq \left(r_1^n - r_2^n + \ldots + r_{2m+1}^n\right)^{1/n}\right\} \quad \text{for a.e. } \Theta \in T \text{ ,}$$

$$D^{*(p)}(\Theta) = \left\{(r,\Theta) \mid \varepsilon \leq r \leq \right.$$

$$\left. \leq \begin{cases} \left(r_1^{n-p} - r_2^{n-p} + \ldots + r_{2m+1}^{n-p}\right)^{1/n-p} & \text{if } p \neq n \\ \dfrac{r_1 \cdot r_3 \cdot \ldots \cdot r_{2m+1}}{r_2 \cdot \ldots \cdot r_{2m}} & \text{if } p = n \end{cases} \right\} \quad \text{for a.e. } \Theta \in T \text{ .}$$

Hence (E') is a consequence of Lemmata 2.21 and 2.18.

Our next result shows the necessity of assumption (E) in Lemma 2.1.

Corollary 2.33

There exists a domain $\Omega \subset \mathbb{R}^2$ and functions u and v which satisfy (A2.5e) and (A2.4) but for which (P1) fails, i.e.

$$\int_\Omega u(x)\, v(x)\, dx > \int_\Omega u^{*(p)}(x)\, v^{*(p)}(x)\, dx \quad \text{holds for } p \geq 1 \text{ .}$$

To prove the corollary let $u = \chi_\Omega$, where $B_3(0) = \Omega$ is a ball with radius 3 and center in the origin and let $\tilde{v} = \chi_{B_\varepsilon(0) \cup B_2(0) \setminus B_1(0))}$. It then follows from the proof of Lemma 2.21, that the support of $\tilde{v}^{*(p)}$ is strictly less than the support of $\tilde{v}^{*(0)}$. Hence there is a mollified version v of $\tilde{v}$ for which (P1) fails.

Property (E') has some interesting consequences. In order to state them we have to define simple functions again.

In this paragraph we call a function $u : \overline{\Omega} \to \mathbb{R}_0^+$ <u>simple</u> if and only if u satisfies (A2.4) and if u is piecewise linear in the sense of affine with respect to the cartesian coordinates $x_1, \ldots, x_{n-1}, x_n$.

In this paragraph we call a function $u : \overline{\Omega} \to \mathbb{R}_0^+$ <u>smooth</u> if and only if $u \in C^1(\Omega)$ satisfies (A2.4) and has the properties: There exists a closed $(n-1)$ dimensional nullset $N \subset T$ such that

i) for every $\theta \in T \setminus N$ and every $c \in \left(\min\limits_{\Omega} u \,,\, \max\limits_{\Omega} u\right)$ the set of points $\{(r,\theta) \in \Omega \mid u(r,\theta) = c\}$ is finite,

ii) for every $\theta \in T \setminus N$ the set of points $\left\{(r,\theta) \in \Omega \,\middle|\, \frac{\partial u}{\partial r}(r,\theta) = 0\right\}$ is finite.

<u>Remark 2.39</u>

In contrast to the monotone decreasing rearrangement in direction y we do not need to define nice functions. If u is simple and if $\frac{\partial u}{\partial r}$ vanishes on a set of positive n-dimensional measure, then ∇u vanishes there a.e., too. In fact if u is piecewise linear, $\frac{\partial u}{\partial r}$ existst a.e. and u is linear (in the sense of affine) on certain polyhedral n-dimension subsets of Ω, or $u \equiv 0$ or $u \equiv$ const on certain subset of Ω.

Let S_q be one of those polyhedrons. If u is not constant in S_q, there is at most one $(n-1)$ dimensional hyperplane H_q such that u is constant on $S_q \cap H_q$ and H_q contains the origin. Here the word hyperplane refers to cartesian coordinates in $\mathbb{R}^n$. Let now M denote the finite union of all such hyperplanes, of the boundaries ∂S_q and of the set in which the Jacobian (2.53) vanishes. Then M is an n-dimensional nullset and $\frac{\partial u}{\partial r} = 0$ implies $\nabla u = 0$ in $\Omega \setminus M$.

Remark 2.40

If u satisfies (A2.4) and $0 \leq u \in W^{1,p}(\Omega)$, $1 \leq p < \infty$, then u can be approximated by simple functions in $W^{1,p}(\Omega)$. Without loss of generality we may assume Ω to have a polyhedral boundary, otherwise we can extend u outside Ω by zero. Then we can approximate u by C_0^∞-functions, which still satisfy (A2.4). This can be done using mollifiers [3, pp. 29, 30, 52; 97, pp. 29/30]. Each C_0^∞-function can then be approximated by linear interpolates as outlined in Remark 2.32.

Now we are able to prove some consequences of (E').

Theorem 2.24

If u satisfies (A2.4), and $0 \leq u \in W^{1,p}(\Omega)$, $1 \leq p < \infty$, then the following inequalities hold:

$$u^{*(o)} \geq u^{*(p)} , \tag{2.63}$$

$$\int_\Omega F(u)\,dz \geq \int_{\Omega^*} F(u^{*(p)}(z))\,dz \quad \text{for } F: \mathbb{R}_0^+ \to \mathbb{R} \text{ continuous and monotone nondecreasing} . \tag{C'}$$

Because of Remark 2.40 it suffices to prove the theorem for simple functions. But for $c > 0$ the level sets of simple functions are polyhedral. We recall the definition of $D(\Theta)$ on p. 13 and see that Lemma 2.22 implies $\Omega_c^{*(p)}(\Theta) \subset \Omega_c^{*(o)}(\Theta)$ for almost every $\Theta \in T$ which proves (2.63) and consequently (C').

In order to prove property (G1) for starshaped rearrangement, we need to express $|\nabla u|$ in terms of n-dimensional spherical coordinates.

Lemma 2.25

Let $x \in \mathbb{R}^n$ be a point in which the Jacobian (2.53) does not vanish. Then under change of variables from cartesian $(x_1,\ldots,x_n)$ to spherical coordinates (r,Θ) we have

$$|\nabla u(x)|^2 = \left|\frac{\partial u}{\partial r}\right|^2 + \sum_{i=1}^{n-1} \frac{k_i(\Theta)}{r^2} \left|\frac{\partial u}{\partial \Theta_i}\right|^2 , \tag{2.64}$$

where $k_i(\Theta)$ are continuous, nonnegative functions.

For the proof we just calculate

$$\frac{\partial u}{\partial x_k} = \frac{\partial u}{\partial r} \cdot \frac{\partial r}{\partial x_k} + \sum_{i=1}^{n-1} \frac{\partial u}{\partial \Theta_i} \cdot \frac{\partial \Theta_i}{\partial x_k} , \tag{2.65}$$

$$\frac{\partial \Theta_i}{\partial x_k} = \frac{r^{n-1}}{j(\Theta)} \cdot \frac{\partial(x_1,\ldots,x_{k-1},x_{k+1},\ldots,x_n)}{\partial(r,\Theta_1,\ldots,\Theta_{i-1},\Theta_{i+1},\ldots,\Theta_{n-1})} \tag{2.66}$$

and observe that the last matrix is $(n-1) \times (n-1)$. Its first column contains functions independent of r and the remaining columns contain expressions of the form r times an angular function. Therefore (2.66) can be rewritten as

$$\frac{\partial \Theta_i}{\partial x_k} := \frac{r^{n-1}}{j(\Theta)} r^{n-2} b_{ik}(\Theta) = \frac{1}{r} \frac{b_{ik}(\Theta)}{j(\Theta)} . \tag{2.67}$$

By assumption $j(\Theta)$ is not zero. Now the orthogonality of the spherical coordinates implies (2.64) with

$$k_i(\Theta) = \frac{1}{j^2(\Theta)} \sum_{k=1}^{n} b_{ik}^2(\Theta) .$$

Theorem 2.26

Suppose that (A2.4) and $\overline{\Omega} = \overline{\Omega}^*$ holds, and that u is simple or smooth. Let $1 \leq p < \infty$ and let $H(t,\Theta)$ be real valued, nonnegative and continuous on $\mathbb{R} \times T$.

i) Then the inequality

$$\int_{\Omega} H(u,\Theta) \, |\nabla u|^p \, dz \geq \int_{\Omega^*} H(u^{*(p)},\Theta) \, |\nabla u^{*(p)}|^p \, dz \text{ holds.} \tag{G1e}$$

ii) Furthermore for $p \geq 1$ and positive H the equality in (G1e) implies $u = u^{*(p)}$.

The proof of (G1e) was given in [21 , Thm. 2.1] for H independent of u or Θ and $p > 1$. For the reader's convenience and since we want to prove ii) we shall essentially repeat and suitably modify their proof. For typographical reasons we shall denote $u^{*(p)}$ by u^* during this proof.

If u is simple we remove the set M which was constructed in Remark 2.39 from the domain of integration. If u is smooth we remove the set where $j(\Theta)$ vanishes and where properties i) and ii) of smooth functions fail, from Ω . Then (G1e) reduces to property

$$\int_{\Omega(\Theta)} H(u,\Theta)\ |\nabla u|^p\ r^{n-1}\ dr \ \geq \int_{\Omega^*(\Theta)} H(u^*,\Theta)\ |\nabla u^*|^p\ r^{n-1}\ dr \qquad (2.68)$$

for almost every $\Theta \in T$. If u is simple the points on $\Omega(\Theta)$ in which u is not differentiable form a finite partitioning of $\Omega(\Theta)$, if u is smooth, so do the points in which $\frac{\partial u}{\partial r}$ vanishes. Let $a_1 \leq a_2 \leq \dots \leq a_M$ be the values assumed by u at these points. Since $\Omega(\Theta)$ does not contain any hyperplane H_q (cf. Remark 2.39), we know that $\frac{\partial u}{\partial r} = 0$ on a nonempty open subinterval of $\Omega(\Theta)$ implies $\nabla u = 0$ on this subinterval, so that $\nabla u^* = 0$ on a corresponding subinterval of $\Omega^*(\Theta)$. Thus we can disregard these sets in the proof of the theorem. Hence it suffices to show

$$\int_{D_k} H\ |\nabla u|^p\ r^{n-1}\ dr \ \geq \int_{D_k^*} H\ |\nabla u^*|^p\ r^{n-1}\ dr \quad \text{for } k = 1, \dots, M-1 \qquad (2.69)$$

where

$D_k := \{r \in \Omega(\Theta) \mid a_k < u(r,\Theta) < a_{k+1}\} \cap CM$ and

$D_k^* := \{r \in \Omega(\Theta) \mid a_k < u^*(r,\Theta) < a_{k+1}\}$ for $k = 1, \dots, M$, and

where CM denotes the complement of M in Ω.

Denote those intervals, at the endpoints of which u assumes the values a_k and a_{k+1} by $\gamma_{k,j}$, $j = 1, \dots, m(\Theta)$.

Let $I_k := \{r \in \Omega(\Theta) \mid a_{k+1} > u^*(r,\Theta) > a_k\}$. Then (2.69) can be rewritten, using Lemma 2.25 as:

$$\sum_{j=1}^{m} \int_{\gamma_{k,j}} H(u,\Theta) \left\{ \left|\frac{\partial u}{\partial r}\right|^2 + \sum_{i=1}^{n-1} \frac{k_i(\Theta)}{r^2} \left|\frac{\partial u}{\partial \Theta_i}\right|^2 \right\}^{p/2} r^{n-1}\ dr \ \geq$$

$$\geq \int_{I_k} H(u^*,\Theta) \left\{ \left|\frac{\partial u^*}{\partial r}\right|^2 + \sum_{i=1}^{n-1} \frac{k_i(\Theta)}{r^2} \left|\frac{\partial u^*}{\partial \Theta_i}\right|^2 \right\}^{p/2} r^{n-1}\ dr \ . \qquad (2.70)$$

The r^{-2} in the curly bracket causes difficulties. We let it disappear and trade in a managable difficulty instead. (2.70) is equivalent to (2.71):

$$\sum_{j=1}^{m} \int_{\gamma_{k,j}} H(u,\Theta) \left\{ r^2 \left|\frac{\partial u}{\partial r}\right|^2 + \sum_{i=1}^{n-1} k_i(\Theta) \left|\frac{\partial u}{\partial \Theta_i}\right|^2 \right\}^{p/2} r^{n-1-p}\, dr \geq$$

$$\geq \int_{I_k} H(u^*,\Theta) \left\{ r^2 \left|\frac{\partial u^*}{\partial r}\right|^2 + \sum_{i=1}^{n-1} k_i(\Theta) \left|\frac{\partial u^*}{\partial \Theta_i}\right|^2 \right\}^{p/2} r^{n-1-p}\, dr \quad . \tag{2.71}$$

The $r^{n-1-p}\,dr$ can be interpreted as $d\rho$, where $\rho = G(r) - G(\varepsilon)$ is a weighted radial variable. This is why $g(r) = r^{n-1-p}$ is so convenient.

A simple calculation shows

$$r \left|\frac{\partial u}{\partial r}\right| = r \cdot \left|\frac{\partial \tilde{u}}{\partial \rho}\right| \left|\frac{d\rho}{dr}\right| = \tilde{p}(\rho) \left|\frac{\partial u}{\partial \rho}\right| \quad ,$$

where $\tilde{u}(\rho,\Theta) = u(r(\rho),\Theta)$, and where

$$\tilde{p}(\rho) := \begin{cases} (n-p)\,\rho + \varepsilon^{n-p} & \text{if } n \neq p \\ 1 & \text{if } n = p \end{cases} \quad , \tag{2.72}$$

so that we can rewrite (2.71) as:

$$\sum_{j=1}^{m} \int_{\tilde{\gamma}_{k,j}} H(\tilde{u},\Theta) \left\{ \tilde{p}(\rho)^2 \left|\frac{\partial \tilde{u}}{\partial \rho}\right|^2 + \sum_{i=1}^{n-1} k_i(\Theta) \left|\frac{\partial \tilde{u}}{\partial \Theta_i}\right|^2 \right\}^{p/2} d\rho \geq$$

$$\geq \int_{\tilde{I}_k} H(\tilde{u}^*,\Theta) \left\{ \tilde{p}(\rho)^2 \left|\frac{\partial \tilde{u}^*}{\partial \rho}\right|^2 + \sum_{i=1}^{n-1} k_i(\Theta) \left|\frac{\partial \tilde{u}^*}{\partial \Theta_i}\right|^2 \right\}^{p/2} d\rho \quad . \tag{2.73}$$

Here $\tilde{\gamma}_{k,j}$ and $\tilde{I}_k$ are the sets $\gamma_{k,j}$ and I_k after the transformation $\rho = G(r) - G(\varepsilon)$. Since $\tilde{u} = u_{max}$ at the origin and $\tilde{u} = u_{min}$ on $\partial\Omega$, there is an odd number m of intervals $\tilde{\gamma}_{k,j}$, $j = 1, \ldots, m$. Suppose that these intervals are ordered by their distance from $\rho = 0$, so that $\tilde{\gamma}_{k,1}$ is the nearest.

Then

$$\text{sign } \frac{\partial \tilde{u}}{\partial \rho} = (-1)^j \quad \text{in } \tilde{\gamma}_{k,j} \quad . \tag{2.74}$$

For fixed k and $\lambda \in (a_k, a_{k+1})$ we denote by $\rho_j(\lambda,\Theta)$ the unique value of ρ in $\tilde{\gamma}_{k,j}$ for which $\tilde{u}(\rho_j,\Theta) = \lambda$ and by $\rho^*(\lambda,\Theta)$ the value of ρ in $\tilde{I}_k$ for which $\tilde{u}^*(\rho^*,\Theta) = \lambda$.

By definition of $\tilde{u}^*$

$$\rho^*(\lambda,\Theta) = \sum_{j=1}^{m} (-1)^{j+1} \rho_j(\lambda,\Theta) \quad . \tag{2.75}$$

Note that (2.75) is considerably simpler than (2.36) because of our strong assumptions on u . Now we replace the variable ρ by λ . Using

$$\frac{\partial \tilde{u}}{\partial \rho} = \left(\frac{\partial \rho_j}{\partial \lambda}\right)^{-1} , \quad \frac{\partial \tilde{u}}{\partial \Theta_i} = - \left(\frac{\partial \rho_j}{\partial \Theta_i}\right) \left(\frac{\partial \rho_j}{\partial \lambda}\right)^{-1} \text{ in } \tilde{\Upsilon}_{k,j} \tag{2.76}$$

as well as (2.74) and (2.75), inequality (2.71) becomes

$$\sum_{j=1}^{m} \int_{a_k}^{a_{k+1}} \left[\tilde{p}^2(\rho_j) + \sum_{i=1}^{n-1} k_i(\Theta) \left|\frac{\partial \rho_j}{\partial \Theta_i}\right|^2\right]^{p/2} \left|\frac{\partial \rho_j}{\partial \lambda}\right|^{1-p} d\lambda \geq$$

$$\geq \int_{a_k}^{a_{k+1}} \left[\tilde{p}^2 \left(\sum_{j=1}^{m} (-1)^{j+1} \rho_j\right) + \sum_{i=1}^{n-1} k_i(\Theta) \left\{\sum_{j=1}^{m} (-1)^{j+1} \frac{\partial \rho_j}{\partial \Theta_i}\right\}^2\right]^{p/2} \times$$

$$\times \left\{\sum_{j=1}^{m} \left|\frac{\partial \rho_j}{\partial \lambda}\right|\right\}^{1-p} d\lambda \ . \tag{2.77}$$

It remains to verify (2.77).

If $p > 1$ an application of Hölder's inequality gives

$$\sum_{j=1}^{m} \left[\tilde{p}^2(\rho_j) + \sum_{i=1}^{n-1} k_i(\Theta) \left|\frac{\partial \rho_j}{\partial \Theta_i}\right|^2\right]^{1/2} =$$

$$= \sum_{j=1}^{m} \left\{\left[\tilde{p}^2(\rho_j) + \sum_{i=1}^{n-1} k_i(\Theta) \left|\frac{\partial \rho_j}{\partial \Theta_i}\right|^2\right]^{1/2} \cdot \left|\frac{\partial \rho_j}{\partial \lambda}\right|^{(1-p)/p}\right\} \left|\frac{\partial \rho_j}{\partial \lambda}\right|^{(p-1)/p} \leq$$

$$\leq \left\{\sum_{j=1}^{m} \frac{\left[\tilde{p}^2(\rho_j) + \sum_{i=1}^{n-1} k_i(\Theta) \left|\frac{\partial \rho_j}{\partial \Theta_i}\right|^2\right]^{p/2}}{\left|\frac{\partial \rho_j}{\partial \lambda}\right|^{p-1}}\right\}^{1/p} \left\{\sum_{j=1}^{m} \left|\frac{\partial \rho_j}{\partial \lambda}\right|\right\}^{(p-1)/p} . \tag{2.78}$$

We interpret the first expression in (2.78) as a sum of l^2-norms of vectors. Then Minkowski's inequality implies

$$\sum_{j=1}^{m} \left[\tilde{p}^2(\rho_j) + \sum_{i=1}^{n-1} k_i(\Theta) \left|\frac{\partial \rho_j}{\partial \Theta_i}\right|^2\right]^{1/2} \geq \left[\left\{\sum_{j=1}^{m} \tilde{p}(\rho_j)\right\}^2 + \right.$$

$$\left. + \sum_{i=1}^{n-1} k_i(\Theta) \left\{\sum_{j=1}^{m} \left|\frac{\partial \rho_j}{\partial \Theta_i}\right|\right\}^2\right]^{1/2} . \qquad (2.79)$$

Now we observe the special structure (2.72) of $\tilde{p}(\rho)$ and claim

$$\sum_{j=1}^{m} \tilde{p}(\rho_j) \geq \tilde{p}\left(\sum_{j=1}^{m} (-1)^{j+1} \rho_j\right) = \tilde{p}(\rho^*) . \qquad (2.80)$$

For $n = p$ this is obvious. For $n > p$, $\rho > 0$ the function $\tilde{p}$ is monotone increasing and positive, which implies

$$\tilde{p}(\rho_1 - \rho_2 + \ldots + \rho_m) \leq \tilde{p}(\rho_m) \leq \sum_{j=1}^{m} \tilde{p}(\rho_j) .$$

For $n < p$ the function $\tilde{p}$ is monotone decreasing and positive, which implies

$$\tilde{p}(\rho_1 - \rho_2 + \ldots + \rho_m) \leq \tilde{p}(\rho_1) \leq \sum_{j=1}^{m} \tilde{p}(\rho_j) .$$

Therefore (2.80) holds and equality holds only if $m = 1$.

Now we raise the inequalities (2.78) and (2.79) to the p-th power and use (2.80) to see that (2.77) does hold. This proves the first part of the theorem for $p > 1$. For $p = 1$ it suffices to use (2.79) and (2.80).

The second statement follows from the fact that equality in (G1e) holds only if $m(k) = 1$ for every $k = 1, \ldots, M(\Theta)$ and almost every $\Theta \in T$.

<u>Corollary 2.27</u>

<u>Suppose that</u> (A2.4) <u>holds and that</u> $0 \leq u \in W_o^{1,p}(\Omega)$. <u>Then the inequality</u>

$$\int_\Omega |\nabla u|^p \, dx \geq \int_{\Omega^*} |\nabla u^{*(p)}|^p \, dx$$ <u>holds for</u> $1 < p < \infty$. (G1e)

The proof of the corollary follows from Theorem 2.26 after extending u by zero onto a larger domain D which satisfies $\overline{D} = \overline{D}^*$ and after approximation by simple functions as described in Remark 2.40 and in the proof of Corollary 2.10.

Example 2.7 "Jets and Cavities", an exterior free boundary problem in potential flow.

There are a few model problems in the calculus of variations, whose solution has brought significant progress to the field. One of them is the obstacle problem which led to the theory of variational inequalities. Another one has recently been solved in the pioneering paper of W. Alt and L.A. Caffarelli [4]. Its physical motivation comes from fluid dynamics and A. Friedman devotes almost one third of his book [75] to this problem. We shall apply starshaped rearrangement to the problem and derive among other things the Lipschitz continuity of the free boundary in a new fashion. Notice that [4] could only derive regularity of the free boundary in "flat" boundary points. Further developments of this problem can be found in [5] and [6].

Let Ω_1 be a compact set in $\mathbb{R}^n$ with boundary $\partial\Omega_1$ of class C^1. Consider the problem

Minimize

$$J_4(v) := \int_{\mathbb{R}^n} \left\{ |\nabla v(x)|^2 + \lambda^2 \chi_{\{v>0\}}(x) \right\} dx \tag{2.81}$$

over

$$\mathbb{K}_4 := \left\{ v \in W^{1,2}_{loc}(\mathbb{R}^n) \mid v \equiv 1 \quad \text{on} \quad \Omega_1 \right\} .$$

Then it is known [4, 75] that there exists a solution u to (2.81). Any solution has bounded support, $0 \le u \le 1$, and belongs to $C^{0,1}(\mathbb{R}^n)$. Furthermore u is harmonic in the set $\{x \in \mathbb{R}^n \setminus \Omega_1 \mid u(x) > 0\} =: D$. Formally u is a solution to the free boundary problem

$$\begin{aligned} \Delta u &= 0 && \text{in } D , \\ u &= 1 && \text{on } \partial\Omega_1 , \\ u = 0 \quad \text{and} \quad \left|\frac{\partial u}{\partial n}\right| &= \lambda && \text{on } \partial\{u > 0\} . \end{aligned} \tag{2.82}$$

Notice that in particular the last boundary condition has to be interpreted in a generalized sense [4, Thm. 2.5], since the boundary $\partial\{u > 0\}$ might not be sufficiently smooth to define a normal vector field on it. For the two dimensional case, however, the free boundary is analytic and for higher dimensions one can define a normal field in every point x from a reduced boundary set $\partial_{red}\{u > 0\} \subset \partial\{u > 0\}$, where $\partial\{u > 0\}\setminus\partial_{red}\{u > 0\}$ is a set of $(n-1)$ dimensional Haussdorff-measure zero. We refer to [4] for details. If the free boundary satisfies the interior sphere condition, which is an open problem for $n \geq 3$, one can follow an argument of D. Tepper and prove the uniqueness of solutions to (2.82) [190; 191; 75, p. 399]. The derivation of the interior sphere condition can also be circumvented as it was conveniently done in [178].

Independent of this uniqueness question we claim:

Corollary 2.28

If $\Omega_1 \supset U_\varepsilon(0)$ and is starshaped with respect to zero, and if u is any solution to problem (2.81) then the support of u and all its level sets Ω_c have to be starshaped with respect to zero. Moreover if Ω_1 is starshaped with respect to each point y in a small nonempty open neighborhood $U_\delta(0)$ of the origin, in particular if Ω_1 is convex, then the free boundary is Lipschitz continuous.

Remark 2.41

For $n = 2$ the starshapedness of $\{u > 0\}$ was derived by D. Tepper by an entirely different method. He also showed that for $n = 2$ the convexity of Ω_1 implies the convexity of $\operatorname{supp} u$.

Remark 2.42

Notice that for $n \geq 3$ the Lipschitz continuity of the free boundary is a new result. H.W. Alt and L. Caffarelli proved regularity for the free boundary only under a flatness condition and showed that there are solutions of (2.82) which have singular boundary. These are, however, only critical points and not global minima of the associated variational functional [4, p. 110].

The second part of the corollary follows from the first in the following way. Let $x \in \{u > 0\}$. Then there exists a cone $C \subset \mathbb{R}^n$ with vertex x, which contains $U_\delta(0)$ such that $\partial\{u > 0\} \cap C = \{x\}$. Otherwise the support of u would not be starshaped with respect to each point in $U_\delta(0)$.

The proof of the first part of Corollary 2.28 is easy in the case that u is unique. Then we can replace u by $u^{*(2)}$ and use Corollary 2.27 and Theorem 2.24 to conclude

$$J_4(u^{*(2)}) \leq J_4(u) \quad . \tag{2.83}$$

Thus $u = u^{*(2)}$ by uniqueness. If u is not known to be unique one has to work a little harder and use

Theorem 2.29

If Ω_1 is compact and u, v are two different solutions to problem (2.81), then they are ordered and their supports are nested, i.e. either $\operatorname{supp} u \subsetneqq \operatorname{supp} v$ and $v > u$ in $\{v > 0\}\setminus\Omega_1$ or $\operatorname{supp} v \subsetneqq \operatorname{supp} u$ and $v < u$ in $\{u > 0\}\setminus\Omega_1$.

Let us postpone the proof of this theorem and continue with the proof of Corollary 2.28. If u is a solution of (2.81) then so is $u^{*(2)}$ and because of Theorem 2.29 the support of u and $u^{*(2)}$ are nested or $u = u^{*(2)}$. If $\operatorname{supp} u \subsetneqq \operatorname{supp} u^{*(2)}$ then we obtain a contradiction to (2.63). If $\operatorname{supp} u \supsetneqq \operatorname{supp} u^{*(2)}$ then $J_4(u) > J_4(u^{*(2)})$, and u cannot be a minimizer of J_4, another contradiction. Thus $u = u^{*(2)}$.

For the proof of Theorem 2.29 we modify an idea which was used by A. Friedman and D. Phillips [77]. It is an easy calculation to see that if u and v are two different solutions of (2.81) then $w_1(x) := \min\{u(x),v(x)\}$ and $w_2(x) := \max\{u(x),v(x)\}$ are solutions of (2.81). In fact

$$J_4(u) = J_4(v) \leq J_4(w_i) \quad \text{for } i = 1, 2 \text{ , but}$$

$$J_4(w_1) + J_4(w_2) \leq J_4(u) + J_4(v) \quad , \text{ so that}$$

$$J_4(v) = J_4(w_1) = J_4(w_2) = J_4(u) \quad .$$

Since w_2 is a solution of problem (2.82) it is harmonic in $(\{u > 0\} \cup \{v > 0\})\setminus\Omega_1$. Therefore

$$\Delta(u-w_2) = 0 = \Delta(v-w_2) \quad \text{in} \quad (\{u > 0\} \cap \{v > 0\})\setminus\Omega_1 = \{w_1 > 0\}\setminus\Omega_1$$

and

$$u - w_2 \leq 0 \quad , \quad v - w_2 \leq 0 \quad \text{on} \quad \partial(\{w_1 > 0\}\backslash\Omega_1) \quad .$$

Now the strong maximum principle implies (either $u < w_2$ or $u \equiv w_2$) and (either $v < w_2$ or $v \equiv w_2$) in $\{w_1 > 0\}\backslash\Omega_1$, so that either $u \equiv v$ or $u < v$ or $v < u$ in $\{w_1 > 0\}\backslash\Omega_1$.

The first case can be ruled out by assumption. In the second case supp $u \subsetneq$ supp v, otherwise v is not continuous. The third case is treated analogously.

<u>Example 2.8</u> Capacitary problems.

Let $\Omega_1 \subset \mathbb{R}^n$ be starshaped with respect to the origin and contain a nonempty open ε-neighborhood of zero. Let $\overline{\Omega}_1 \subset\subset \Omega_o \subset \mathbb{R}^n$, where Ω_o is bounded and starshaped with respect to zero. Let $u \in C^{0,1}(\overline{\Omega}_o)$, $0 \leq u \leq 1$ be a Lipschitz continuous solution of the variational problem.

Minimize

$$J_5(v) \quad := \int_{\Omega_1} \left\{\frac{1}{p} \, |\nabla v(x)|^p + F(v(x))\right\} dx \qquad (2.84)$$

over

$$\mathbb{K}_5 \quad := \left\{v \in W_o^{1,p}(\Omega_o) \mid v \equiv 1 \quad \text{on} \quad \Omega_1\right\} \quad ,$$

where $F : [0,1] \to \mathbb{R}$ is continuous and monotone nondecreasing and $p > 1$. If F is differentiable with derivative $f : [0,1] \to \mathbb{R}$, then the solutions u of the variational problem (2.84) are weak solutions of the degenerate elliptic boundary value problem

$$\operatorname{div}\,(|\nabla u|^{p-2}\,\nabla u) \;=\; f(u) \quad \text{in} \quad \Omega_o\backslash\overline{\Omega}_1 \;, \qquad (2.85)$$
$$u \;=\; 1 \quad \text{on} \quad \partial\Omega_1 \;, \quad u \;=\; 0 \quad \text{on} \quad \partial\Omega_o \;.$$

For the regularity of solutions to such problems we refer e.g. too [32, 68, 171, 193]. If F is convex with subdifferential ∂F, then u is the unique weak solution of the differential inclusion

$$\operatorname{div}\,(|\nabla u|^{p-2}\,\nabla u) \;\in\; \partial F(u) \qquad (2.86)$$

under the boundary conditions of (2.85) (cf. e.g. [61,100].

Corollary 2.30

In addition to the above assumptions suppose that at least one of the conditions i) ii) or iii) holds:

i) *Problem (2.84) has a unique solution.*

ii) *Problem (2.84) has only smooth solutions.*

iii) *If problem (2.84) has two different solutions* u *and* w , *then either* $\operatorname{supp} u \supset \operatorname{supp} w$ *and* $u > w$ *in* $\operatorname{supp} u \setminus \overline{\Omega}_1$, *or* $\operatorname{supp} w \supset \operatorname{supp} u$ *and* $w > u$ *in* $\operatorname{supp} w \setminus \overline{\Omega}_1$. *Furthermore let* F *be nonconstant.*

Then every solution u *of problem (2.84) has level sets which are star-shaped with respect to the origin.*

The proof of Corollary 2.30 is almost trivial under assumptions i) or ii) since $J_5(u^{*(p)}) < J(u)$ unless $u = u^{*(p)}$ as a consequence of Theorem 2.24, 2.26 and Corollary 2.27. Under assumption iii) we argue as follows: Theorem 2.24 implies

$$\int_\Omega F(u)\,dx = \int_\Omega F(u^{*(o)})\,dx \geq \int_\Omega F(u^{*(p)})\,dx ,$$

so that $u^{*(p)} < u$ in $\operatorname{supp} u \setminus \overline{\Omega}_1$ unless $u = u^{*(p)}$. But then again $J_5(u^{*(p)}) < J_5(u)$.

Remark 2.43

Previous versions of Corollary 2.30 for special F and p (mainly $F \equiv 0$, $p = 2$) are known [20, 21, 93, 96, 130, 152]. The novelty is that they can now be derived without monotonicity assumptions on f and without differentiability assumptions on f and u . Our corollary applies for instance to the nondifferentiable function $f_1(u) = (u - \frac{1}{2})^+$ or to $p = 2$ and $f_2(u) = \sqrt{u}\,(1-u)$, since then u is known to be unique. A case in which assumption iii) holds is [77]:

$$p = 2 \quad \text{and} \quad f(t) = \begin{cases} t^\alpha f_o(t) & \text{for } t \geq 0 \\ 0 & \text{for } t < 0 \end{cases} \quad \text{for some } \alpha \in (0,1)$$

and $m \leq f_o(t) \leq M$, $0 < m \leq M < \infty$, $f_o \in C^2(\mathbb{R})$.

In [125] M. Longinetti gives a quantitative characterization of star-shapedness for $n = p = 2$.

Example 2.9 Obstacle problems.

Let $\Omega_o \subset \mathbb{R}^n$ be starshaped with respect to zero and let $\partial\Omega_o$ be sufficiently smooth. Let $\psi \in C^{1,1}(\Omega_o)$ be given with $\psi < 0$ on $\partial\Omega_o$, $\psi(0) = \psi_{max} > 0$ and suppose that all the level sets of ψ are starshaped with respect to zero.

Corollary 2.31

Let u be the solution of the variational problem $(1 < p < \infty)$.

Minimize

$$J_6(v) := \int_{\Omega_o} |\nabla u(x)|^p \, dx \tag{2.87}$$

over

$$\mathbb{K}_6 := \left\{ v \in W_o^{1,p}(\Omega_o) \mid v \geq \psi \quad \text{a.e. in } \Omega_o \right\} .$$

Then all the level sets of u are starshaped with respect to zero. This can be deduced from Theorem 2.26 and property (M1) by cutting off ψ at "height" $\psi(0) - \delta$ and by the limiting process $\delta \to 0^+$ [96]. For $p = 2$ another proof of this result was already given in [93] under the slightly stronger assumption $x \cdot \nabla\psi < 0$ in $\Omega_o \backslash \{0\}$.

Remark 2.44a

There are many variational inequalities which can be reduced to obstacle problems, e.g. the dam problem (Example 2.6). We already know that the free boundary in the dam problem has to be starshaped with respect to the origin. If we cut off the solution u of the dam problem at "height" $\frac{1}{2} H^2 - \delta$, where $\delta > 0$ is sufficiently small, we can deduce the Lipschitz continuity of the free boundary as in Corollary 2.28. We omit the details, since the result is already known.

Remark 2.44b

Some results on starshapedness and convexity of the coincidence set $\{x \in \Omega \mid u(x) = \psi(x)\}$ can be found in [77, 99, 105b, 161]. Note that equimeasurable rearrangements such as Steiner and Schwarz symmetrization can also be applied to obstacle problems [146], and even to obstacle problems with volume constraints as treated by G. Eisen [66].

Example 2.10

Exterior free boundary problems in the context of reaction diffusion.

Let Ω_1 be given as in Example 2.8 and let $u : \mathbb{R}^n \to \mathbb{R}_0^+$ be a Lipschitz continuous solution of the variational problem

Minimize

$$J_7(v) \quad := \int_{\mathbb{R}^n \setminus \Omega_1} \left\{ \frac{1}{p} |\nabla v|^p + \lambda^2 F(v) \right\} dx \qquad (2.88)$$

over

$$\mathbb{K}_7 \quad := \quad \left\{ v \in W^{1,p}_{loc}(\mathbb{R}^n) \mid v \equiv 1 \quad \text{on} \quad \Omega_1 \right\} \quad .$$

Here $F : [0,1] \to \mathbb{R}$ is assumed convex, lower semicontinuous and monotone nondecreasing and $1 < p < \infty$. Furthermore let $0 \in \partial F(0)$ and suppose $\int_0^1 [F(t)]^{-1/p} dt$ is finite. Then one can see from a standard comparison argument that the support of the unique solution u of (2.88) has to be bounded. This was kindly pointed out to me by J.I. Diaz [61]. Furthermore u is a weak solution of

$$\begin{aligned} \operatorname{div}(|\nabla u|^{p-2} \nabla u) \quad &\in \quad \lambda^2 f(u) \quad \text{in} \quad \mathbb{R}^n \setminus \bar{\Omega}_1 \quad , \\ u \quad &= \quad 1 \qquad \text{on} \quad \Omega_1 \quad , \end{aligned} \qquad (2.89)$$

where $f = \partial F$.

As λ tends to infinity, the support of u decreases and shrinks to Ω_1 and a boundary layer. Phenomena of this type were investigated (for $p = 2$) by L.S. Frank and W.D. Wendt [74] for the mapping

$$f_3(u) \quad := \quad \begin{cases} \{1\} & \text{if} \quad u > 0 \quad , \\ [0,1] & \text{if} \quad u = 0 \quad , \\ \{0\} & \text{if} \quad u < 0 \quad , \end{cases}$$

and by A. Friedman and D. Phillips [77] for a class of functions which includes the mapping

$$f_4(u) \quad := \quad \{(u^+)^\alpha\} \quad \text{with} \quad 0 < \alpha < 1 \quad .$$

As in Corollaries 2.28 and 2.30 we can now deduce that the starshapedness of Ω_1 with respect to zero implies starshapedness of supp u and that the starshapedness of Ω_1 with respect to $U_\varepsilon(0)$ implies

Lipschitz continuity of the free boundary $\partial\{x \in \mathbb{R}^n | u(x) > 0\}$. To this end we approximate u by a minimizing sequence of simple functions, rearrange each element of the sequence and use the weak lower semicontinuity of J_7 in $W_o^{1,p}(B_R(0))$ where $B_R(0)$ is a sufficiently large ball containing supp u .

In [100] it was shown that the convexity of Ω_1 implies the convexity of the support and all level sets of u , provided f is monotone nondecreasing, see also § III.11 of these notes.

II.7 Steiner symmetrization with respect to $\{y = 0\}$

We use the notation introduced in the definition of Steiner symmetrization in § II.1. In particular let us recall that a point $z \in \mathbb{R}^n$ is denoted by $(x',y) \in \mathbb{R}^{n-1} \times \mathbb{R}$.

We shall prove a generalization of property (G2f) because this generalization will imply property (G2h) for circular symmetrization as a corollary. In this paragraph we shall use the assumptions:

$$\Omega = \Omega' \times (-\omega,\omega) \text{ , where } \Omega' \subset \mathbb{R}^{n-1} \text{ is a bounded domain} \qquad \text{(A2.5f)}$$

and

$$u : \overline{\Omega} \to \mathbb{R}_o^+ \text{ is Lipschitz continuous and } u = 0 \text{ on } \partial\Omega \text{ ;} \qquad \text{(A2.6f)}$$

or (A2.5f) and

$$u : \overline{\Omega} \to \mathbb{R} \text{ is Lipschitz continuous and periodic in } y \text{ , i.e. } u(x',\omega) = u(x',-\omega) \text{ for } x' \in \Omega' \text{ .} \qquad \text{(A2.7f)}$$

Under these assumptions one can easily modify the proofs of Lemmata 2.8 and 2.12 and see that u^* is again Lipschitz continuous. Since the proof does not contain any new ideas, we omit the details. Again we define simple, nice and smooth functions.

In this paragraph we call a function $u : \overline{\Omega} \to \mathbb{R}$ <u>simple</u> if and only if Ω satisfies (A2.5f), if $u \in C(\overline{\Omega})$ and if u is piecewise linear in the sense of affine.

We call a function $u : \overline{\Omega} \to \mathbb{R}$ nice if and only if it is simple and if $\frac{\partial u}{\partial y} \neq 0$ a.e. in .

In this paragraph we call a function $u : \overline{\Omega} \to \mathbb{R}$ smooth if and only if $u \in C^1(\overline{\Omega})$, if Ω satisfies (A2.5f) and if u has the following properties:

There exists a closed subset $N \subset \overline{\Omega}'$ of $(n-1)$ dimensional measure zero such that

i) for every $x' \in \Omega'\setminus N$ and every $c \in \left(\min_{\Omega} u \, , \, \max_{\Omega} u\right)$ the set of points $\{(x',y) \in \Omega | \; u(x',y) = c\}$ is finite,

ii) for every $x' \in \Omega'\setminus N$ the set of points $\left\{(x',y) \in \Omega \middle| \frac{\partial u}{\partial y}(x',y) = 0\right\}$ is finite.

Remark 2.44c

If Ω' is polyhedral, nice functions are dense in $W^{1,p}(\Omega)$, $1 < p < \infty$. We refer to Remark 2.32.

Theorem 2.31

Let Ω satisfy (A2.5f).

Let $u : \overline{\Omega} \to \mathbb{R}$ be nice or smooth and satisfy a) (A2.6f) or b) (A2.7f).

Let $F : \Omega' \times \mathbb{R} \to \mathbb{R}_o^+$ and $X_k : \Omega' \to \mathbb{R}_o^+$ $(k = 1,\dots,n)$ be nonnegative and continuous and let $G : \mathbb{R}_o^+ \to \mathbb{R}$ be monotone nondecreasing and convex

i) Then the inequality

$$\int_{\Omega} F(x',u) \; G\left(\left\{\sum_{k=1}^{n-1} X_k(x') \left|\frac{\partial u}{\partial x_k}\right|^2 + X_n(x') \left|\frac{\partial u}{\partial y}\right|^2\right\}^{1/2}\right) dx' \, dy \geq$$

$$\geq \int_{\Omega} F(x',u^*) \; G\left(\left\{\sum_{k=1}^{n-1} X_k(x') \left|\frac{\partial u^*}{\partial x_k}\right|^2 + X_k(x') \left|\frac{\partial u^*}{\partial y}\right|^2\right\}^{1/2}\right) dx' \, dy \quad . \tag{G2f}$$

holds.

ii) If moreover F and X_k $(k = 1,\dots,n)$ are positive and G is monotone increasing and strictly convex, then equality holds in (G2f)

a) if and only if $u = u^*$, under assumption (A2.6f), or

b) <u>if and only if</u> $u = u^*$ <u>modulo translation, under periodicity assumptions</u> (A2.7f).

For the proof we proceed essentially the same way as in the proof of Theorems 2.9 and 2.13.

The proof is modelled after the one of G. Polya and G. Szegö [56, pp. 154], who gave it for $n = 3$, $F \equiv 1$, $X_k \equiv 1$, $k = 1, 2, 3$, and for "sufficiently smooth" functions. Notice that in [156, p. 186] the question of discussing the equality sign was dismissed as hopeless. We have to show that for almost every $x' \in \Omega'$ the following inequality holds:

$$\int_{-\omega}^{\omega} F(x',u)\, G(\bullet)\, dy \;\geq\; \int_{-\omega}^{\omega} F(x',u^*)\, G(\bullet)\, dy \quad . \tag{2.90}$$

After removing a closed set N of $(n-1)$ dimensional Lebesgue measure zero from Ω' we may assume the following: If u is nice then for every $x' \in \Omega' \backslash N$, $u(x',y)$ is differentiable with respect to all variables for all but finitely many $y \in (-\omega,\omega)$.

If u is smooth there at most are finitely many points on the segment $\{(x',y)\ ,\ y \in (-\omega,\omega)\}$ in which $\frac{\partial u}{\partial y}(x',y) = 0$. Suppose there are M such exceptional points and let $a_1 \leq a_2 \leq \dots \leq a_M$ be the values assumed by u at these M points. Define

$$D_i = \{y \in \Omega(x') \mid a_i < u(x',y) < a_{i+1}\}$$

and

$$D_i^* = \{y \in \Omega(x') \mid a_i < u^*(x',y) < a_{i+1}\} \quad \text{for } i = 1,\dots,M-1 .$$

It suffices to integrate over these sets in (2.90). Now fix i and decompose D_i into a finite number of intervals $\{\gamma_{i,j}\}_{j=1,\dots,N(i,x')}$ in each of which u is differentiable (with respect to all variables) and $\frac{\partial u}{\partial y} \neq 0$. For each $\lambda \in (a_i, a_{i+1})$ denote by $y_j(\lambda,x')$ the unique value of y in $\gamma_{i,j}$ for which $u(x',y_j) = \lambda$. The function $u^*(x',\bullet)$ is monotone decreasing in y on $[0,\omega]$.

Denote by $y^*(x',\lambda)$ the unique nonnegative value of y in D_i^* for which $u^*(x',y^*(x',\lambda)) = \lambda$. We assume that the intervals $\{\gamma_{i,j}\}_{j=1,\dots,N}$ are ordered by their distance from $-\omega$, so that by definition

$$\operatorname{sign} \frac{\partial u}{\partial y}(y_j,x') = (-1)^{j+1} \operatorname{sign} \frac{\partial u}{\partial y}(y_1,x') \quad \text{in} \quad \Upsilon_{i,j} , \tag{2.35}$$

and

$$y^*(\lambda,x') = \frac{1}{2} \begin{cases} \sum_{j=1}^{N} (-1)^j \, y_j(\lambda,x') & \text{if } \frac{\partial y_1}{\partial \lambda} > 0 \text{ and } \frac{\partial y_N}{\partial \lambda} < 0 , \\ \sum_{j=1}^{N} (-1)^{j+1} \, y_j(\lambda,x') & \text{if } \frac{\partial y_1}{\partial \lambda} < 0 \text{ and } \frac{\partial y_N}{\partial \lambda} > 0 , \end{cases} \tag{2.91}$$

hold. Notice that N has to be even because of (A2.6f) or (A2.7f).

In each interval $\Upsilon_{i,j}$ the function $\frac{\partial u}{\partial y}$ is nonzero, and hence by the implicit function theorem $y_j(\lambda,x')$ is differentiable. Thus we obtain

$$\frac{\partial u}{\partial y} = \left(\frac{\partial y_j}{\partial \lambda}\right)^{-1} \quad \text{in} \quad \Upsilon_{i,j} \tag{2.37}$$

$$\frac{\partial u}{\partial x_k} = -\frac{\partial y_j}{\partial x_k}\left(\frac{\partial y_j}{\partial \lambda}\right)^{-1} \quad \text{for} \quad k = 1,\dots,n-1 \quad \text{in} \quad \Upsilon_{i,j} \tag{2.38}$$

and

$$\frac{\partial u^*}{\partial y} = \left(\frac{\partial y^*}{\partial \lambda}\right)^{-1} \tag{2.40}$$

$$\frac{\partial u^*}{\partial x_k} = -\left(\frac{\partial y^*}{\partial x_k}\right)\left(\frac{\partial y^*}{\partial \lambda}\right)^{-1} . \tag{2.41}$$

Furthermore (2.91) implies the relations:

$$\left|\frac{\partial y^*}{\partial \lambda}\right| = \frac{1}{2} \sum_{j=1}^{N} \left|\frac{\partial y_j}{\partial \lambda}\right| \quad \text{and} \quad \left|\frac{\partial y^*}{\partial x_k}\right| = \frac{1}{2} \left|\sum_{j=1}^{N} (-1)^j \frac{\partial y_j}{\partial x_k}\right| . \tag{2.92}$$

Now we can rewrite the modified gradients, the arguments of G, in terms of derivatives with respect to (x',λ) as follows:

$$\sum_{k=1}^{n-1} X_k(x') \left|\frac{\partial u}{\partial x_k}\right|^2 + X_n(x') \left|\frac{\partial u}{\partial y}\right|^2 = \left|\frac{\partial y_j}{\partial \lambda}\right|^{-2} \left\{ \sum_{k=1}^{n-1} X_k(x') \left|\frac{\partial y_j}{\partial x_k}\right|^2 + X_n(x') \right\} \quad \text{in} \ \Upsilon_{i,j} \tag{2.93}$$

$$\sum_{k=1}^{n-1} X_k(x') \left|\frac{\partial u^*}{\partial x_k}\right|^2 + X_n(x') \left|\frac{\partial u^*}{\partial y}\right|^2 = \left|\frac{\partial y^*}{\partial \lambda}\right|^{-2} \left\{\sum_{k=1}^{n-1} X_k(x') \left|\frac{\partial y^*}{\partial x_k}\right|^2 + \right.$$

$$\left. + X_n(x')\right\} \quad \text{in} \quad D_i^* . \qquad (2.94)$$

In order to prove (2.90) we have to show

$$\sum_{k=1}^{N} \int_{\gamma_{i,j}} F(x',u) \; G\left(\left\{\sum_{k=1}^{n-1} X_k(x') \left|\frac{\partial u}{\partial x_k}\right|^2 + X_n(x') \left|\frac{\partial u}{\partial y}\right|^2\right\}^{1/2}\right) dy \geq$$

$$\geq 2 \int_{D_i^* \cap \{(x',y)\,|\,y \geq 0\}} F(x',u^*) \; G\left(\left\{\sum_{k=1}^{n-1} X_k(x') \left|\frac{\partial u^*}{\partial x_k}\right|^2 + \right.\right.$$

$$\left.\left. + X_n(x') \left|\frac{\partial u^*}{\partial y}\right|^2\right\}^{1/2}\right) dy \quad . \qquad (2.95)$$

for every $i = 1, \ldots, M-1$. Let us fix i and introduce λ as variable of integration. Then it remains to show

$$\sum_{j=1}^{N} \left\{\left|\frac{\partial y_j}{\partial \lambda}\right| G\left(\left|\frac{\partial y_j}{\partial \lambda}\right|^{-1} \left\{\sum_{k=1}^{n-1} X_k(x') \left|\frac{\partial y_j}{\partial x_k}\right|^2 + X_n(x')\right\}^{1/2}\right)\right\} \geq$$

$$\geq \left(\sum_{j=1}^{N} \left|\frac{\partial y_j}{\partial \lambda}\right|\right) G\left[2\left(\sum_{j=1}^{N} \left|\frac{\partial y_j}{\partial \lambda}\right|\right)^{-2} \left\{\frac{1}{4} \sum_{k=1}^{n-1} X_k(x') \left|\sum_{j=1}^{N} (-1)^j \frac{\partial y_j}{\partial x_k}\right|^2 + \right.\right.$$

$$\left.\left. + X_n(x')\right\}^{1/2}\right) . \qquad (2.96)$$

Again we set $\alpha_j := \left|\frac{\partial y_j}{\partial \lambda}\right| \left(\sum_{j=1}^{N} \left|\frac{\partial y_j}{\partial \lambda}\right|\right)^{-1}$ and observe $\sum_{j=1}^{N} \alpha_j = 1$, so that (2.96) reduces to showing

$$\sum_{j=1}^{N} \left\{\sum_{k=1}^{n-1} X_k(x') \left|\frac{\partial y_j}{\partial x_k}\right|^2 + X_n(x')\right\}^{1/2} \geq \left\{\sum_{k=1}^{n-1} \left(X_k(x') \left|\sum_{j=1}^{N} (-1)^j \frac{\partial y_j}{\partial x_k}\right|^2\right) + \right.$$

$$\left. + 4\, X_n(x')\right\}^{1/2} , \qquad (2.97)$$

since G is assumed to be nondecreasing and convex. If we introduce the notation $a_o^j := +\sqrt{X_n(x')}$ for $j = 1, \ldots, N$; $a_k^j := +\sqrt{X_k(x')} \cdot \left|\frac{\partial y_j}{\partial x_k}\right|$ for $k = 1, \ldots, n-1$ and $j = 1, \ldots, N$, then the left hand side of (2.97) represents the sum (over j) of the length of the n-vectors $a^j := (a_o^j,\dots,a_{n-1}^i)$. We apply Minkowski's inequality and verify (2.97):

$$\sum_{j=1}^{N} \left\{X_n(x') \sum_{k=1}^{n-1} X_k(x') \left|\frac{\partial y_j}{\partial x_k}\right|^2\right\}^{1/2} = \sum_{j=1}^{N} \left(\sum_{k=0}^{n-1} (a_k^j)^2\right)^{1/2} \geq$$

$$\geq \left\{\sum_{k=0}^{n-1} \left(\sum_{j=1}^{N} a_k^j\right)^2\right\}^{1/2}$$

$$= \left\{X_n(x') \left(\sum_{j=1}^{N} 1\right)^2 + \sum_{k=1}^{n-1} \left(\sqrt{X_k(x')} \sum_{j=1}^{N} \left|\frac{\partial y_j}{\partial x_k}\right|\right)^2\right\}^{1/2}$$

$$\geq \left\{4\, X_n(x') + \sum_{k=1}^{n-1} X_k(x') \left|\sum_{j=1}^{N} (-1)^j \frac{\partial y_j}{\partial x_k}\right|^2\right\}^{1/2} . \tag{2.98}$$

This proves part i) of the theorem. An inspection of the proof yields that under the strengthened assumptions on F, X_k and G equality holds only under the following conditions: $N(i,x) = 2$ for $i = 1, \dots, M-1$ and almost every $x' \in \Omega'$; $\frac{\partial y_1}{\partial x_k} = - \frac{\partial y_2}{\partial x_k}$ and $\frac{\partial y_1}{\partial \lambda} = - \frac{\partial y_2}{\partial \lambda}$ for almost every $x' \in \Omega'$ and λ . This implies ii).

<u>Corollary 2.32</u>

<u>Let</u> $\Omega \subset \mathbb{R}^n$ <u>be a bounded domain, and let</u> $0 \leq u \in W_o^{1,p}(\Omega)$, $1 < p < \infty$. <u>Then</u> $u^* \in W_o^{1,p}(\Omega^*)$, <u>and the inequality</u>

$$\int_\Omega F(x',u)\, |\nabla u|^p\, dx'\, dy \geq \int_{\Omega^*} F(x',u^*)\, |\nabla u^*|^p\, dx'\, dy \tag{2.99}$$

<u>holds under the assumptions of Theorem</u> 2.31 i) <u>on</u> F .

Here Ω^* is by definition the interior of $\overline{\Omega}^*$. For the proof one extends u by zero on $Q\setminus\Omega$, where Q is a sufficiently large hypercube. So without loss of generality we may assume that Ω satisfies (A2.5f). Then one approximates u by nice functions u_n . The sequence $\{u_n\}_{n\in\mathbb{N}}$ has a weakly convergent subsequence in $W^{1,p}(Q)$. The rest of the proof is identical to the one of Corollary 2.14.

<u>Remark 2.45</u>

As a simple application of Steiner symmetrization let us recall Example 2.5. Let u be a solution of this problem. Now suppose that $u^*(x,y)$ is the symmetric decreasing rearrangement of u with respect to

$\{x = 0\}$. Then Corollary 2.32, property (C) and Theorem 2.1 imply $J_2(u^*) \leq J_2(u)$. Since $u^* \in \mathbb{K}_2$ and u is unique, it follows $u = u^*$. This gives another proof of Corollary 2.17.

Remark 2.46

In [181] Steiner symmetrization was used for a compactness argument. Other results that were obtained through Steiner symmetrization are described in the book of G. Polya and G. Szegö and in [100, 126, 127, 192].

Remark 2.47

Another conceivable way to "symmetrize" a function $u(x',y)$ with respect to $\{y = 0\}$ is to replace it by $\tilde{u}(x',y) = \frac{1}{2}(u(x',y) + u(x',-y))$. This approach has some shortcomings. It can only be used for minima of convex variational problems, which are in general unique, and therefore symmetric. And it does not reveal whether u is symmetrically decreasing in y .

Open problem

If $-\Delta u = f$ in Ω , $u = 0$ on $\partial\Omega$, where Ω is bounded and $0 \leq f \in C^\infty(\Omega)$ and if $-\Delta v = f^*$ in Ω^* , $v = 0$ on $\partial\Omega^*$, show that $\|u\|_{L^\infty(\Omega)} \leq \|v\|_{L^\infty(\Omega^*)}$. This result is known to hold for Schwarz symmetrization and circular and spherical symmetrization, but apparently unkown for Steiner symmetrization.

Remark 2.48

Nonequimeasurable symmetric decreasing rearrangements can be found in [80, 151].

II.8 Schwarz symmetrization

This is the most frequently used kind of symmetrization. We prefer the name Schwarz symmetrization, since the notion of spherically symmetric rearrangement can easily be confused with spherical symmetrization (h). Recent results on Schwarz symmetrization can be found in the books [19, 140] of C. Bandle and J. Mossino, historical remarks and classical results are described in G. Polya and G. Szegö's notes [156].

The literature on Schwarz symmetrization is growing rapidly and should not be ignored in these notes.

Hence we shall briefly report on some progress of others in this area and then give some of our own results. One of the first powerful applications of Schwarz symmetrization was the proof of the Krahn Faber inequality [69, 113]: Among all fixed membranes of given area, the circular one has the lowest principal eigenvalue.

This was shown by looking at

$$\lambda_1(\Omega) = \operatorname*{Min}_{\substack{u \in \overset{\circ}{W}{}^{1,2}(\Omega) \\ u \neq 0}} \frac{\int_\Omega |\nabla u|^2 \, dx}{\int_\Omega u^2 \, dx} .$$

(G1) and (C) imply $\lambda_1(\Omega) \geq \lambda_1(\Omega^*)$. Polya and Szegö proved a famous conjecture of Saint Venant, by looking at its variational formulation

$$P(\Omega) = \operatorname*{Max}_{v \in W_0^{1,2}(\Omega)} \frac{4\left(\int_\Omega v \, dx\right)^2}{\int_\Omega |\nabla v|^2 \, dx} .$$

Among all bars of given area of cross section Ω , the circular cross section yields maximal torsional rigidity. Only recently M.T. Kohler-Jobin showed the following:

Among all plane domains of given torsional rigidity P the circle has the lowest eigenvalue. This result is not quite so "trivial". Obviously $P(\Omega) \leq P(\Omega^*)$ and $\lambda_1(\Omega) \geq \lambda_1(\Omega^*)$.

But if we replace Ω^* by a smaller circle Ω_1^* such that $P(\Omega) = P(\Omega_1^*)$, $\lambda_1(\Omega^*) \leq \lambda_1(\Omega_1^*)$ and one has to prove $\lambda_1(\Omega_1^*) \leq \lambda_1(\Omega)$.

A nonequimeasurable version of Schwarz symmetrization can be found in [103] where each level set Ω_c is replaced by a circle of same α-mean radius as Ω_c . A famous result concerns the <u>best constant in Sobolev's inequality</u> [117, 132, 184]. We want to find a constant K depending only on the "size" $m_n(\Omega)$ of Ω such that

$$\|u\|_{L^q(\Omega)} \leq K \|\nabla u\|_{L^p(\Omega)}$$

for $u \in W_o^{1,p}(\Omega)$, $\Omega \subset \mathbb{R}^n$, $1 < p < n$ and $q < \frac{n\,p}{n-p}$. For $p = q = 2$ such a constant is provided by $\frac{1}{\lambda_1(\Omega^*)} \left(\geq \frac{1}{\lambda_1(\Omega)}\right)$. For other p, q we just minimize

$$R(u) = \frac{\|\nabla u\|_{L^p(\Omega)}}{\|u\|_{L^q(\Omega)}}$$

over $W_o^{1,p}(\Omega)$.

Clearly $R(u) \geq R(u^*)$ where u^* is in $W_o^{1,p}(\Omega^*)$. So in order to find the best constant it suffices to consider the symmetrized problem. This general phenomenon, that certain estimates become optimal in the symmetrized case, can be observed again and again.

Another case concerns the problems

$$\begin{aligned} -\Delta u(x) &= f(x) \text{ in } \Omega , & -\Delta v(x) &= f^*(x) \text{ in } \Omega^* , \\ u(x) &= 0 \text{ on } \partial\Omega , & v(x) &= 0 \text{ on } \partial\Omega^* , \end{aligned}$$

where one can show [19, p. 176; 183] $\|v\|_{L^\infty(\Omega^*)} \geq \|u^*\|_{L^\infty(\Omega^*)} = \|u\|_{L^\infty(\Omega)}$, provided $\Omega \subset {}^n$ is a bounded domain and e.g. $f \in L^\infty(\Omega)$. Similar statements can be made for the minimal solution of $\Delta u + f(u) = 0$ in Ω and $\Delta v + f(v) = 0$ in Ω^* under homogeneous Dirichlet conditions, provided $f \in C^\alpha(\mathbb{R})$, $f(0) > 0$ and f is strictly monotone increasing. Then $\|u\|_{L^p} = \|u^*\|_{L^p} \leq \|v\|_{L^p}$ for any $p \in (1,\infty]$ [19, p. 189].

In the same spirit one can vary f, λ and L in the elliptic problem $L\,u = f$ in Ω and compare a solution u of

$$\begin{aligned} L\,u &= f \text{ in } \Omega , \\ u &= 0 \text{ on } \partial\Omega , \end{aligned}$$

with a solution v of

$$L^* v = f^* \quad \text{in } \Omega ,$$

$$v = 0 \quad \text{on } \partial\Omega ,$$

where e.g. $L u = - \sum_{i,j=1} a_{ij}(x) \frac{\partial^2 u}{\partial x_i \, \partial x_j}$ and $L^* u = - \lambda \Delta u$.

Here λ is the ellipticity constant of L . This was first done by H. Weinberger [199] and later generalized up to degenerate quasilinear operators L in [83, 184, 188], see also [7] for the strictness of this comparison.

There are also investigations to prove similar results for nonlinear parabolic equations and they contain some beautiful results [19, p. 216; 82; 195, 196]. Finally one has applied Schwarz symmetrization to get a priori bounds for solutions of Hamilton-Jacobi equations [84].

To demonstrate how actively this field of research is investigated, let us furthermore refer to [8-11, 13, 14, 24, 25, 27-29, 38a, 40-42, 51-53, 65, 89, 105a, 128, 133, 146, 195, 196].

After quoting these impressive results, let us make two new contributions. First we answer a conjecture of E. Lieb [116, p. 97]. We call a domain $\Omega \subset \mathbb{R}^n$ smooth if $\partial\Omega$ is piecewise analytic.

Corollary 2.33

Let $\Omega \subset \mathbb{R}^n$ be a bounded smooth domain and $u : \overline{\Omega} \to \mathbb{R}_0^+$ analytic in Ω with zero boundary values. Let $1 < p < \infty$. Then the strict equality in

$$\int_\Omega |\nabla u|^p \, dx \geq \int_{\Omega^*} |\nabla u^*|^p \, dx \tag{G2g}$$

holds only if u equals u^* modulo translation.

For various proofs of inequality (G2g) we refer to [19, 81, 91, 116, 156, 174, 184]. The proof of the corollary will use only arguments from Steiner symmetrization. Without loss of generality we may assume that u attains its maximum in the origin, otherwise we translate u . If u is not equal to u^* , then there exists a $n-1$ dimensional hyperplane through the origin such that u is not symmetrically decreasing with respect to this hyperplane. If we denote a point in $\mathbb{R}^n$ by $z = (x',y) \in \mathbb{R}^{n-1} \times \mathbb{R}$, we can assume that $\{y = 0\}$ is this hyperplane. Now an inspection and slight modification of the proof of Theorem 2.31

reveals that the Steiner symmetrization $\tilde{u}$ of u with respect to y (strictly) decreases the L^p-norm of $|\nabla u|$. Finally consider the Schwarz symmetrization $\tilde{u}^*$ of $\tilde{u}$, which obviously coincides with u^* . We arrive at the strict inequality

$$\int_\Omega |\nabla u|^p \, dz > \int_{\tilde{\Omega}} |\nabla \tilde{u}|^p \, dz \geq \int_{\Omega^*} |\nabla u^*|^p \, dz \quad .$$

Another proof of this corollary would follow from an inspection of [19, p. 55]. Recently still another proof of the corollary was indicated by A. Friedman and B. McLeod in [76, Thm. 2.2].

Remark 2.49

Corollary 2.33 holds neither for arbitrary positive functions in $W_o^{1,p}(\Omega)$ (cf. Example 2.3), nor for $p = 1$. For $p = 1$ we give the following (nonanalytic) counter-example. Suppose a nonnegative function v is defined on the unit ball in $\mathbb{R}^2$, has vanishing boundary data and attains its positive maximum in a point which is not the origin. Furthermore suppose that the level lines of v , i.e. the sets $\partial\{v(x) \geq c\}$ are nested (excentric) circles. Then we know from the Fleming Rishel formula

$$\int_{\Omega^*} |\nabla v^*| \, dz = \int_\Omega |\nabla v| \, dz$$

while $v^* \neq v$.

Example 2.11 The sharpness of the Krahn-Faber inequality.

As mentioned above, Krahn and Faber proved the conjecture of Lord Rayleigh that $\lambda_1(\Omega) \geq \lambda_1(\Omega^*)$, where λ_1 is the first eigenvalue of the Laplace operator under Dirichlet conditions. As noted in [165] and stated in [69, 113] they also showed that equality holds only if Ω equals Ω^* modulo translation.

This sharpness result is not recorded in the standard references [156] and [148] and there have been other attempts to prove it. A heuristic argument was provided by G. Polya in [155]. The proof of L. Tonelli [194] is incomplete since it only leads to the conclusion that each level set Ω_c has to be a circle, and in the preceding remark we constructed an example of a function v whose level sets are circles, but which does not coincide with its Schwarz symmetrization. Also the proof in [19, p. 104] contains a small gap: the criticism in Remark 2.23 applies to it. Both proofs can be easily completed by using the positivity and analyticity of the first eigenfunction.

Our Corollary 2.33 gives an independent proof of the fact that among all smooth domains Ω of same n-dimensional Lebesgue measure equality holds only if Ω equals Ω^* modulo translation.

Example 2.12 A problem of A. Friedman.

Consider the variational problem $(\lambda > 0)$:

Minimize

$$J_8(v) := \int_{B_1} \left\{ |\nabla v(x)| + \lambda \chi_{\{v>0\}}(x) \right\} dx \tag{2.100}$$

over

$$\mathbb{K}_8 := \left\{ v \in W^{1,2}(B_1) \mid v \equiv 1 \text{ on } \partial B_1 \right\} .$$

Here B_1 denotes the unit ball in $\mathbb{R}^n$. H.W. Alt and L.A. Caffarelli proved that problem (2.100) has Lipschitz continuous solutions. In his book [75, p. 283] A. Friedman proposed to prove the following result from which one can infer uniqueness.

Corollary 2.34

If u is a solution of problem (2.100), then u has rotational symmetry, i.e. $u(x) = u(|x|)$.

The proof is not as trivial as suggested in [75]. Let $u_*(x) := -(-u(x))^*$ be the spherically symmetric increasing rearrangement of u . If u is a solution to problem (2.100), then so is u_* . This does not automatically mean $u = u_*$, since u and u_* could be different solutions. To close this gap one has to prove that different solutions of (2.100) are nested in the sense of Theorem 2.29, and this can be done exactly as in the proof of Theorem 2.29.

So if the dead core $D = \{x \in B_1 \mid u(x) = 0\}$ is nonempty and $u = u^*$, we get a contradiction to Theorem 2.29 and the equimeasurability of D and D* . And if the dead core is empty, then $u \equiv 1$ by uniqueness.

Remark 2.50

Another way to transform functions $u(x)$ into rotationally symmetric functions are the spherical means $\bar{u}$ of u. Here $\bar{u}(r)$ is the mean value of u over a sphere of radius r with center in the origin. In general u and $\bar{u}$ are not equimeasurable. In a different context spherical means can be useful tools for a priori estimates, too, because they reduce partial differential equations to ordinary ones (cf. e.g. [197]).

Open problem

Under what assumptions on $u : \bar{\Omega} \to \mathbb{R}_0^+$ can one generate the Schwarz symmetrization u^* as a limit of successive Steiner symmetrizations of u? For $u(x) = \chi_D(x)$ and D a bounded measurable set this has been shown by Brascamp, Lieb and Luttinger, and it seems that a combination of their proof and our proof of Lemma 2.1 should give the desired result for Lebesgue measurable functions.

II.9 Circular and spherical symmetrization

In this paragraph we want to point out that circular symmetrization is really just a variant of Steiner symmetrization. The notation u^* will now refer to the circular symmetrization of u. To avoid technicalities we shall not state the results under optimal assumptions. The following results were previously known: If $0 \le u \in W_0^{1,p}(\Omega)$ and if $\Omega \subset \mathbb{R}^2$ is a bounded smooth domain, and if u is sufficiently smooth, then the inequalities [89, p. 76 ff.; 156, p. 194; 175]

$$\int_\Omega |\nabla u|^p \, dx \ge \int_{\Omega^*} |\nabla u^*|^p \, dx \quad \text{for} \quad 1 \le p < \infty \qquad \text{(G1h)}$$

and

$$\int_\Omega \sqrt{1 + |\nabla u|^2} \, dx \ge \int_{\Omega^*} \sqrt{1 + |\nabla u^*|^2} \, dx$$

hold.

Here we prove a stronger result and discuss the equality sign.

Corollary 2.35

i) If $\Omega \subset \mathbb{R}^2$ is a bounded domain, $F : \mathbb{R} \times \mathbb{R} \to \mathbb{R}_0^+$ continuous, if $1 < p < \infty$ and if $0 \leq u \in W_0^{1,p}(\Omega)$, then $u^* \in W_0^{1,p}(\Omega^*)$ and the inequality (G1h) holds:

$$\int_\Omega F(|x|,u)\ |\nabla u(x)|^p\ dx \geq \int_{\Omega^*} F(|x|,u^*)\ |\nabla u^*(x)|^p\ dx\ . \qquad \text{(G1h)}$$

ii) If in addition to the assumptions of i) the domain Ω is smooth, i.e. $\partial\Omega$ piecewise analytic, and if the function u is analytic in Ω, and if $G : \mathbb{R}_0^+ \to \mathbb{R}$ is monotone nondecreasing and convex, then inequality (G2h) holds:

$$\int_\Omega F(|x|,u)\ G(|\nabla u|)\ dx \geq \int_{\Omega^*} F(|x|,u^*)\ G(|\nabla u^*|)\ dx\ . \qquad \text{(G2h)}$$

iii) If in addition to the assumptions of ii) the function F is continuous and positive and G increasing and strictly convex, and if Ω is an annulus, $\Omega = \{x \in \mathbb{R}^2 | 0 \leq a < |x| < b\}$, then equality holds in (G2h) if and only if $u = u^*$ modulo rotation.

For the proof we just visualize Ω as a domain in the (r,φ)-plane. In case i) we extend u by zero on a sufficiently large rectangle in the (r,φ)-plane and apply Corollary 2.32. Notice that in polar coordinates we can identify y with φ, x with r, X_n with $1/r$ and X_1 with r. In case iii) we note that either $\frac{\partial u}{\partial \varphi} \equiv 0$, in which case the proof is trivial or u is smooth in the sense required by Theorem 2.31. Finally ii) follows from the observation that Theorem 2.31 holds even if Ω does not satisfy (A2.5f) but if Ω is smooth and if $u \geq 0$ vanishes on $\partial\Omega$.

Let us give two applications of this corollary.

Example 2.13 Nonlinear eigenfunctions.

Consider the problem

$$\Delta u - u + u^{2N+1} = 0\ , \quad u \geq 0 \quad \text{in } \Omega \subset \mathbb{R}^2 \qquad (2.101)$$

$$u = 0 \qquad \text{on } \partial\Omega\ , \qquad (2.102)$$

where $\Omega := \{z \in \mathbb{R}^2 | 0 < a < |z| < b\}$ is an annulus and $N \in \mathbb{N}$. What can be said about the geometry of solutions to this problem? We introduce some notation. For $k \in \mathbb{N}$ let V_k denote the positive cone in the subspace of $W_o^{1,2}(\Omega)$ which is invariant under rotation by $\frac{2\pi}{k}$;

$$V_k := \left\{ v \in W_o^{1,2}(\Omega) \,|\, v(r,\varphi) = v\left(r,\varphi + \frac{2\pi}{k}\right) , \ v > 0 \ \text{ a.e. in } \ \Omega \right\} .$$

Therefore a function $v \in V_k$ is at least k-times periodic with respect to φ in $[0,2\pi]$.

In consistence with the definition of V_k let V_∞ denote the set of rotationally invariant positive functions

$$V_\infty := \left\{ v \in W_o^{1,2}(\Omega) \,|\, v(r,\varphi) = v(r) \ , \ v > 0 \ \text{ a.e. in } \ \Omega \right\} .$$

If the width $b - a$ of the annulus is prescribed, then it was shown by C. Coffman [55] that problem (2.101) (2.102) has solutions u_k in $V_k \backslash V_\infty$, $1 \le k \le k_o(a)$ and $\lim_{a\to\infty} k_o(a) = \infty$.

Notice that these solutions depend explicitly on φ and are at least k-times periodic in φ . We intend to show that they are precisely k-times periodic in φ . This will be a (rare) situation in which symmetrization increases the multiplicity of solutions.

The solutions u_k of problem (2.101) (2.102) can be characterized in terms of a variational problem. Consider the associated Rayleigh quotient

$$R_2(v) := \frac{\left[\int_\Omega (|\nabla v|^2 + v^2)\, dx\right]^{N+1}}{\int_\Omega v^{2N+2}\, dx}$$

and let ω_k be a solution of the minimization problem

$$\min_{v\in V_k} R_2(v) \quad , \text{ i.e. } \quad R_2(v) \ge R_2(\omega_k) \quad \text{for} \quad v \in V_k \ .$$

Such a solution is known to exist. Then ω_k satisfies the Euler equation

$$\Delta\omega_k - \omega_k + \lambda\, \omega_k^{2N+1} = 0 \quad \text{in } \Omega \ , \qquad (2.103)$$

where

$$\lambda = R_2(\omega_k) \left\|\omega_k\right\|^{-2N}_{W_o^{1,2}(\Omega)} \ .$$

Now we can set $u_k = \lambda^{\frac{1}{2N}} \omega_k$ and verify that u_k satisfies (2.101). The fact that u_k belongs to V_k does not imply much about the oscillatory behavior of $u_k(r,\varphi)$ for fixed r. In fact u_k could be nk-times periodic in φ, $n \in \mathbb{N}$.

If we use the additional information that ω_k satisfies a variational problem, however, we can show

$$u_k \in V_k \setminus \left\{V_\infty \cup \bigcup_{n=2}^{\infty} V_{nk}\right\} .$$

This is in fact a consequence of the following result.

Corollary 2.36

For $1 < n \in \mathbb{N}$ and $k \in \mathbb{N}$ the strict inequality $R_2(\omega_k) < R_2(\omega_{nk})$ holds.

For the proof consider the restriction of ω_{nk} to the annular sector $\overset{k}{\Omega} := \{x = (r,\varphi) \in \Omega \| \varphi| \leq \frac{\pi}{k}, a < r < b\}$. Visualize $\overset{k}{\Omega}$ as domain in the (r,φ)-plane. Replace $\omega_{nk}\big|_{\overset{k}{\Omega}}$ by its Steiner symmetrization $\omega_k^*\big|_{\overset{k}{\Omega}}$ on $\overset{k}{\Omega}$ with respect to φ and extend the symmetrized function periodically to Ω. Let us denote this extension by ω_k^*. Then $\omega_k^* \in V_k$ and because of Theorem 2.31 ii) and property (C) $R_2(\omega_k^*) < R_2(\omega_{nk})$.

Example 2.14 The plasma problem.

Let $\Omega \subset \mathbb{R}^2$ be a bounded domain with boundary of class $C^{2,\gamma}$. A simple model for the shape at equilibrium of a confined plasma leads to a boundary value problem of the following type. For given positive numbers λ and I find a function $u \in C^{2,\gamma}(\bar{\Omega})$ satisfying

$$\Delta u + \lambda\, u^+(x) = 0 \quad \text{in } \Omega , \tag{2.104}$$

$$u(x) = c \quad \text{(unknown constant) on } \partial\Omega , \tag{2.105}$$

$$\int_{\partial\Omega} \left|\frac{\partial u}{\partial n}\right| ds = I . \tag{2.106}$$

The nonlinearity λu^+ in (2.104) could be replaced by more general nonlinearities [22, 23, 102], but for simplicity let us stick to this special case which was first studied by R. Temam. For bibliographical remarks we refer the reader to the book of A. Friedman, where one can also find most of the following facts about problem (2.104) - (2.106). If λ is larger than the smallest eigenvalue of the Laplacian on Ω under Dirichlet boundary conditions, then the constant c in (2.105) is negative, so that a free boundary $\partial\Omega_o := \{x \in \Omega \mid u(x) = 0\}$ occurs. If λ is larger than the second eigenvalue and Ω is not convex the solutions of (2.104) - (2.106) are in general not unique. A sufficient condition for u to be a solution of (2.104) - (2.106) is that u is the minimizer of an appropriate variational problem. Various variational problems have been suggested, some of which are known to be equivalent to each other. A commonly used variational problem is the one by R. Temam:

Minimize

$$J_9(v) := \int_\Omega (|\nabla v|^2 - \lambda(v^+)^2)\, dx + 2\, I\, c \qquad (2.107)$$

over

$$\mathbb{K}_9 := \left\{ v \in W_o^{1,2}(\Omega) \mid v = c \quad \text{on} \quad \partial\Omega\, , \quad \int_\Omega v^+\, dx = \frac{I}{\lambda} \right\} .$$

Solutions of (2.107) are called <u>variational solutions of the plasma problem</u>. It is known that for those solutions the free boundary is a simple closed analytic curve. What is known about their uniqueness?

If Ω is a circle, Schwarz symmetrization shows that they have to be unique. If Ω is convex they are believed to be unique and one can show asymptotic uniqueness for $\lambda \to \infty$ [99, 172]. If Ω is an annulus in $\mathbb{R}^2$, they are not unique, as was first shown by D. Schaeffer [162]. In particular it is known that variational solutions do depend on the angle and thus there are infinitely many of them. Incidentally, if one applies Steiner symmetrization to the solutions described in [163] one can prove symmetry properties of those solutions as well.

<u>Corollary 2.37</u>

<u>If Ω is an annulus and u is a variational solution of the plasma problem, then $u = u^*$ modulo rotation and the function $\frac{\partial u}{\partial \varphi}$ has precisely two nodal domains.</u>

This follows from Theorem 2.31 ii), since u is smooth.

Remark 2.51a

Very recently A. Friedman and B. McLeod have extended the special case $F \equiv 1$, $G(t) = t^p$, $1 < p < \infty$ of Corollary 2.35 to domains in $\mathbb{R}^n$, in particular n-dimensional annuli. Here u^* denotes the spherical symmetrization of u .

This implies that some nonradial positive solutions of $-\Delta u = u^p$ in $\Omega = \{x \in \mathbb{R}^n \mid 0 < a < |x| < b\}$, $u = 0$ on $\partial\Omega$, with $p < \frac{n+2}{n-2}$ and $n \geq 2$ have to coincide with their spherical symmetrizations u^* modulo rotation. It is known that for p close to $\frac{n+2}{n-2}$ there are both nonradial and radial solutions, i.e. some which depend on the angle and some which don't [38b, p. 453].

Remark 2.51b

Spherical symmetrization can be used to derive a priori estimates for solutions of partial differential equations on irregular domains. This was done by E. Sperner [177]. For a similar result we also refer to A. Weitsman [200].

III. MAXIMUM PRINCIPLES

It is the purpose of the subsequent paragraphs to demonstrate that maximum principles can be used to obtain information on level sets of solutions to elliptic boundary value problems. In order to simplify the exposition, we shall restrict our attention to nonnegative solutions of semilinear differential equations. For the existence question for those equations we refer to the survey [119] of P.L. Lions.

III.10 The moving plane method

The moving plane method was invented by Alexandroff, later used by J. Serring [169] and some of its major applications were discoverend by B. Gidas, W.M. Ni, and L. Nirenberg [85]. It has subsequently and successfully been apllied to other problems: to the Stefan problem [136], to parabolic reaction-diffusion equations [92] and even to Hamilton Jacobi equations [30].

We are going to describe two of the main results of [85] and point out two slight extensions of them.

Let $\Omega \subset \mathbb{R}^n$ be bounded domain with boundary of class $C^{2,\alpha}$, $0 < \alpha < 1$, and let u be a classical solution of

$$\begin{aligned} \Delta u + f(u) &= 0 \quad \text{in } \Omega \, , \quad u > 0 \quad \text{in } \Omega \, , \\ u &= 0 \quad \text{on } \partial\Omega \, , \end{aligned} \tag{3.1}$$

where for simplicity f is assumed to be Lipschitz continuous. We denote a point in $\mathbb{R}^n$ by (x',y) with $x' \in \mathbb{R}^{n-1}$.

Theorem 3.1 [85]

If Ω is symmetric and convex in y, then any classical solution u of (3.1) is symmetric in y and $y \cdot \frac{\partial u}{\partial y} < 0$ for $y \neq 0$. In particular if Ω is a ball, with center in the origin, then u depends only on the distance r to the origin and $\frac{\partial u}{\partial r} < 0$ for $r \neq 0$.

Corollary 3.2

If Ω is symmetric and convex in y, then all the level sets Ω_c of u are symmetric and convex in y. Moreover if Ω is symmetric and convex in n orthogonal directions, then all the level sets Ω_c of u are symmetric and convex in those directions, and they are star-shaped. Finally under those assumptions the gradient of u vanishes only in a single point, in the center of symmetry.

Remark 3.1

The number and location of critical points of u is an interesting mathematical subject in itself. Whenever the gradient of u does not vanish, we can use the implicit function theorem to derive regularity of the "free boundary" $\partial\Omega_c$. It is therefore desirable to locate critical points of u. In two dimensions there are results of L. Payne and R. Sperb in this direction [149, 173], an extension is Corollary 3.2 (see also [94]). Another reason why the existence of a single critical point is of interest, was kindly pointed out to the author by K. Schmitt. Associated with problem (3.1) is the so-called P-function $P = |\nabla u|^2 + 2\,F(u)$, where F is the primitive of f. If u is a C^3-solution to problem (3.1) and if Ω is convex, one can show that the function P attains its maximum only in a critical point of u [174]. If there is only one such point, x_o say, then $f(u(x_o))$ has to be nonnegative. This allows one to derive conclusions about the L^∞-norm of ∇u.

Other attempts to bound classical solutions of (3.1) in terms of geometric quantities related to Ω and structural constants of f can be found in [72]. Gidas, Ni and Nirenberg showed that solutions of (3.1) cannot have critical points near the boundary. To explain their result we have to introduce some notation.

Consider the plane $T_\mu := \{z = (x,y) \in \mathbb{R}^n \mid y = \mu\}$. For large positive μ the plane T_μ is disjoint from Ω. As we diminish μ, the plane T_μ will cut off from Ω an open cap $\Sigma(\mu) := \{z \in \Omega \mid y > \mu\}$. Let $\Sigma'(\mu)$ denote the reflection of $\Sigma(\mu)$ across the plane T_μ. In the beginning $\Sigma'(\mu)$ will be contained in Ω, and there will be a smallest μ such that the reflected cap $\Sigma'(\mu)$ is contained in Ω. We set $\mu_1 := \inf\{\mu \mid \Sigma'(\mu) \subset \Omega\}$ and call $\Sigma(\mu_1)$ the optimal cap in direction y of Ω.

Theorem 3.3 [85]

If $\Sigma(\mu_1)$ is the optimal cap in direction y of Ω and if u is a classical solution of (3.1), then $\frac{\partial u}{\partial y} < 0$ in $\Sigma(\mu_1)$.

We want to apply Theorem 3.3 to a problem, which does not have classical solutions, because f is discontinuous. To avoid technicalities, we consider a particular example.

Example 3.1 An interior "dead core" problem and a weak version of [85]. Let $\Omega \subset \mathbb{R}^n$ be a bounded domain with boundary $\partial\Omega$ of class C^2, let $\lambda > 0$ and let u be the solution to the following variational problem:

Minimize

$$J_{10}(v) := \int_\Omega \left\{\frac{1}{2} |\nabla v|^2 + \lambda v^+\right\} dx \tag{3.2}$$

over

$$\mathbb{K}_{10} := \{v \in H^1(\Omega) \mid v = 1 \text{ on } \partial\Omega\} .$$

The associated Euler equation can be written as

$$-\Delta u + \lambda H(u) \ni 0 , \tag{3.3}$$

where $H(u)$ is the multivalved Heaviside function defined by

$$H(t) = \begin{cases} \{1\} & \text{if } t > 0 , \\ [0,1] & \text{if } t = 0 , \\ \{0\} & \text{if } t < 0 . \end{cases} \tag{3.4}$$

It is known [74] that problem (3.2) has a unique solution $0 \leq u \leq 1$, $u \in W^{2,p}(\Omega) \cap C^{1,\alpha}(\Omega)$ for $p \geq 1$, $\alpha < 1$, and that for sufficiently large λ the solution will have a "dead core", i.e. a region in which it is zero. The name "dead core" comes from chemical engineering, where u represents the stationary concentration of a reacting and diffusing chemical in a reactor [27]. Now we can state a consequence of Theorem 3.3. Notice that because of the dead core we cannot expect the conclusion of Theorem 3.3 to hold for $1 - u$. Instead we have

Corollary 3.4

If $\Sigma(\mu_1)$ is the optical cap in direction y of Ω and if u is a solution of (3.2), then $\frac{\partial u}{\partial y} \geq 0$ in $\Sigma(\mu_1)$.

For the proof of the corollary one has to approximate H by its Yosida approximation [39] H_ε and consider the family of problems

$$
\begin{aligned}
-\Delta u_\varepsilon + \lambda H_\varepsilon(u_\varepsilon) &= 0 \quad \text{in } \Omega \quad , \\
u &= 1 \quad \text{on } \partial\Omega \quad .
\end{aligned}
\tag{3.5}
$$

Then it is known that for each $\varepsilon > 0$ there exists a unique classical solution u_ε of problem (3.5) and that the family $\{u_\varepsilon\}_{\varepsilon>0}$ has a subsequence which converges pointwise a.e. to u . To each of these $1 - u_\varepsilon$ we can apply Theorem 3.3 and conclude that $1 - u_\varepsilon$ is nondecreasing in y in the optimal cap $\Sigma(\mu_1)$. This property is preserved under pointwise convergence, and, using the regularity of u , the corollary follows.

Remark 3.2

In the case of two dimensions, A. Friedman and D. Phillips [77] were able to show that convex Ω implies a convex dead core. In higher dimensions this problem was solved in [101].

III.11 Convexity of level sets

Consider again problem (3.1) and suppose that Ω is convex. Then one might conjecture that all the level sets Ω_c of u are convex. This conjecture has puzzled a lot of people [1, 2, 37, 45, 46, 96, 104, 112, 115, 120, 124, 129, 148, 187] and even for $f(u) = \lambda_1 u$ the proof is by no means trivial. There are several strategies to attack this conjecture.

Strategy 1

Show that u is concave on Ω or that $g(u)$ is concave on Ω, where g is a monotone increasing function of its argument. This strategy will be pursued in the next paragraph.

Strategy 2

Show that the principal curvatures of $\partial\Omega_c$ have the right sign. This approach seems to work well in two dimensions [2, 45, 124, 150, 187].

Strategy 3

In two space dimensions show that the Gauss curvature of u or of $g(u)$ is positive in Ω, where g is a monotone increasing function of its argument. This approach was pursued in [95, 186].

Strategy 4

Show that the level sets Ω_c of u are convex, or equivalently show that the inequality

$$u\left(tx + (1-t)\,y\right) \geq \min\{u(x), u(y)\} \tag{3.6}$$

holds for every $t \in [0,1]$ and x and $y \in \overline{\Omega}$. This characterization of quasiconcave functions is well known in optimization theory [164], and we shall give a successful example of this strategy below.

Strategy 5

A variational approach: For the plasma confinement problem, Example 2.14, one can give a different variational formulation which forces the plasma region to be convex. This was done by A. Acker, L. Caffarelli and J. Spruck [1, 46].

Strategy 6

A rearrangement approach: Replace each level set Ω_c of u by a convex level set, e.g. by the convex hull of Ω_c . This way one can define a new function $\tilde{u}$ with convex level sets, a suitable name for $\tilde{u}$ might be quasiconcave envelopes. If u has been characterized as minimum of a variational functional, show that $\tilde{u}$ is a minimum, too. This approach has been tried without success. One reason is that it is hard to estimate the Dirichlet integral of $\tilde{u}$. The observation that convex level sets can be established mainly for solutions of variational problems was one of the motivations for the author to study rearrangement methods.

Another strategy could be to show $u = \tilde{u}$ via maximum principles. To the authors knowledge this has not been done yet. Finally we should mention that M. Longinetti [123] has given a mean-value-type characterization of quasiconcave functions which might be useful for further studies.

Let us now pursue Strategy 4 for a particular example. The method that we use is a variant of the "Gabriel-Lewis" method [100, 115]; the new aspect is that it is applied to a free boundary problem whose solutions are not $C^2(\mathbb{R}^n\backslash\bar{\Omega})$.

Example 3.2

An exterior free boundary problem (cf. Example 2.10).

Let $\Omega_1 \subset \mathbb{R}^n$ be a bounded convex domain with boundary of class $C^{2,\alpha}$. Let $f : \mathbb{R} \to \mathbb{R}$ be a mapping satisfying

$$f \text{ is monotone nondecreasing, } f(t) = 0 \text{ for } t \leq 0 \text{ , } f(t) > 0 \text{ for } 0 < t \leq 1 \text{ .} \tag{3.7}$$

$$f \in C^{1,\beta}((0,1)) \text{ for some } \beta \in (0,1) \quad . \tag{3.8}$$

$$\int_0^1 F(t)^{-1/2}\,dt < \infty \text{ , where } F(t) = \int_0^t f(\tau)\,d\tau \quad . \tag{3.9}$$

Notice that under these assumptions f might have a jump discontinuity at zero and that (3.9) is for instance satisfied if $f(t) \sim t^\alpha$ as $t \downarrow 0$, $0 < \alpha < 1$.

Consider the boundary value problem

$$\Delta u = f(u) \quad \text{in} \quad \mathbb{R}^n \setminus \bar{\Omega}_1 \quad , \tag{3.10}$$

$$u \equiv 1 \quad \text{on} \quad \bar{\Omega}_1 \, , \quad u(x) \to 0 \quad \text{for} \quad |x| \to \infty \quad . \tag{3.11}$$

Here equation (3.10) has to be understood in the weak sense, i.e.

$$-\int_{\mathbb{R}^n} \nabla u \cdot \nabla \varphi \, dx = \int_{\mathbb{R}^n} f(u) \, \varphi \, dx \quad \text{for every} \quad \varphi \in C_0^\infty(\mathbb{R}^n \setminus \bar{\Omega}_1) \quad .$$

It is known that there exists a unique solution $u \in C^{3,\beta}(D) \cap C^{1,\lambda}(\mathbb{R}^n \setminus \bar{\Omega}_1)$ to the exterior problem $0 < \lambda < 1$. Here $D := \{x \in \mathbb{R}^n \setminus \bar{\Omega}_1 \mid 0 < u(x) < 1\}$.

<u>Theorem 3.5</u>

<u>Under the above assumptions on</u> Ω_1 <u>and</u> f <u>all the level sets</u> $\Omega_c := \{x \in \mathbb{R}^n \mid u(x) \geq c\}$ <u>of the solution</u> u <u>to problem</u> (3.10) (3.11) <u>and in particular the support of</u> u <u>are convex</u>.

Aim of the proof is to verify (3.6) or, since u is continuous and the binary numbers are dense in the reals, to verify the inequality

$$Q(x_1,x_2) := u\left(\frac{1}{2} \, x_1+x_2)\right) - \min\{u(x_1),u(x_2)\} \geq 0 \quad \text{in } \mathbb{R}^{2n} \quad . \tag{3.12}$$

Then u is quasiconcave. Let us denote by C the open convex hull of D . Once we manage to show that $C = D$ we can apply arguments from [46, 93] to conclude that other level sets of u are convex. If (3.12) is violated, then there exists a pair of points $(y_1,y_2) \in \bar{C} \times \bar{C}$ such that Q attains a global negative minimum in (y_1,y_2)

$$u\left(\frac{1}{2} \, (x_y+y_2)\right) < \min\{u(y_1),u(y_2)\} \quad , \tag{3.13}$$

and since $u \geq 0$ in C and $u = 0$ on ∂C we already know:

$$\text{Both } y_1 \text{ and } y_2 \quad \left(\text{and} \quad \frac{y_1+y_2}{2}\right) \quad \text{are in} \quad C \quad . \tag{3.14}$$

The next steps in the proof are aimed at showing that y_1, y_2 and $\frac{y_1 + y_2}{2}$ are all in D (Lemmata 3.7b) and 3.9b)). In [93] one can find a proof of the following result. It follows from the differential inequality

$$\Delta\,[(x-x^0) \, \nabla u(x)] \geq f'(u) \, (x-x^0) \quad \nabla u(x) \quad .$$

Lemma 3.6

If supp u and $\overline{\Omega}_1$ are starshaped with respect to $x^o \in \overline{\Omega}_1$, then $(x-x^o)\ \nabla u(x) < 0$ in D.

Using the results of Example 2.10 and the convexity of Ω_1 this implies

$$(x-x^o)\ \nabla u(x) \quad < \quad 0 \quad \text{in} \quad D \quad \text{for any} \quad x^o \in \Omega_1 \ , \qquad (3.15)$$

and in particular

$$|\nabla u(x)| \quad > \quad 0 \quad \text{in} \quad D \ . \qquad (3.16)$$

Lemma 3.7

a) $u(y_1) \quad = \quad u(y_2) > 0$,

b) $y_1 \in D$, $y_2 \in D$ and $\dfrac{y_1 + y_2}{2} \in C\backslash\overline{\Omega}_1$.

Proof of Lemma 3.7a)

Suppose that $u(y_1) < u(y_2)$. Then locally near (y_1,y_2) the quasiconcavity function Q has the C^2-representation

$$Q(x_1,x_2) \quad = \quad u\left(\frac{1}{2}\ (x_1+x_2)\right) - u(x_1)$$

and the gradients of Q with respect to x_1 and x_2 have to vanish at (y_1,y_2) , i.e.,

$$\frac{1}{2}\ \nabla u\left(\frac{1}{2}\ (y_1+y_2)\right) \quad = \quad \nabla u(y_1) \quad = \quad 0 \ .$$

But this means $u(y_1) = 0$ because of (3.16) and hence $Q(y_1,y_2) \geq 0$ in contrast to (3.13). Therefore $u(y_1)$ has to be equal to $u(y_2)$. If they are both zero, then again $Q(y_1,y_2) \geq 0$, a contradiction to (3.13).

Proof of Lemma 3.7b)

We know from Lemma 3.7a) that $y_1 \in D \cup \overline{\Omega}_1$ and $y_2 \in D \cup \overline{\Omega}_1$ and we have to exclude the possibilities i) y_1 and $y_2 \in \overline{\Omega}_1$ and ii)

$y_1 \in \bar{\Omega}_1$ and $y_2 \in D$. In case i) $\frac{y_1 + y_2}{2} \in \bar{\Omega}_1$ by convexity of $\bar{\Omega}_1$ and $Q(y_1, y_2) \geq 0$. In case ii) we apply Lemma 3.6 with $x^o = y_1$ and $x = y_2$ and get even $Q(y_1, y_2) > 0$. Finally, $\frac{y_1 + y_2}{2} \in \bar{\Omega}_1$ implies $Q(y_1, y_2) \geq 0$ since $u \leq 1$ on $\mathbb{R}^n$.

In order to prove that $\frac{y_1 + y_2}{2} \in D$ we have to compare the gradient of u in y_1, y_2 and $\frac{y_1 + y_2}{2}$. Recall that due to the regularity of u and Lemma 3.7b) the function u is differentiable in those three points.

Lemma 3.8

$\nabla u(y_1)$ is parallel and points in the same direction as $\nabla u(y_2)$.

Proof: If this were not the case then there would exist a unit vector $\xi \in \mathbb{R}^n$ such that $u_\xi(y_1) < 0$ and $u_\xi(y_2) > 0$. This would contradtict the fact that Q attains it minimum in (y_1, y_2) since one could diminish Q by moving y_1 in direction $-\xi$ and y_2 in direction ξ.

Lemma 3.9

a) $\nabla u\left(\frac{y_1+y_2}{2}\right)$ does not vanish and is parallel to and equally directed as $\nabla u(y_1)$ or $\nabla u(y_2)$.

b) $\frac{y_1 + y_2}{2} \in D$.

If h is a unit vector in $\mathbb{R}^n$ and t a small positive number, then

$$Q\,(y_1 + th,\, y_2 + th) - Q(y_1, y_2) \geq 0 .$$

Therefore $\nabla u\left(\frac{y_1+y_2}{2}\right)$ does not vanish and has to point in the same direction as $\nabla u(y_1)$. The second assertion follows from the first and from (3.16).

Let us summarize that up to this point we have ruled out many possible locations of the extremal triple y_1, y_2 and $\frac{y_1 + y_2}{2}$. The remaining alternative that y_1, y_2 and $\frac{y_1 + y_2}{2}$ are all in D can be excluded because of the special structure of equation (3.10). Notice that u is three times differentiable in a neighborhood of y_1, y_2 and $\frac{y_1 + y_2}{2}$, so that we can use equation (3.10). This was in fact sketched in [93] and for the readers convenience we shall work out the details here. In the first part of the following lemma we improve the qualitative statements of Lemmata 3.8 and 3.9 to a quantitative one. We introduce the notation

$$a \;:=\; \left|\nabla u\left(\frac{y_1+y_2}{2}\right)\right| \;,\quad b \;:=\; |\nabla u(y_1)| \quad\text{and}\quad c \;:=\; |\nabla u(y_2)| \;.$$

Lemma 3.10

a) $\frac{1}{a} = \frac{1}{2}\left(\frac{1}{b} + \frac{1}{c}\right)$

b) $\frac{1}{a^2}\,\Delta u\left(\frac{y_1+y_2}{2}\right) \;\geq\; \frac{\mu}{b^2}\,\Delta u(y_1) + \frac{(1-\mu)}{c^2}\,\Delta u(y_2)$, where $\mu = \frac{c}{b+c} \in (0,1)$.

For the proof let us define $n := \frac{1}{b}\,\nabla u(y_1)$ and let us fix a unit vector $h \in \mathbb{R}^n$ with the property

$$h \cdot n \neq 0 \quad . \tag{3.17}$$

We want to move from y_1 to $y_1 + \frac{s}{b}\,h$ and from y_2 to $y_2 + \frac{r(s)}{c}\,h$ in such a way that the side constraint $u(x_1) = u(x_2)$ remains satisfied. Notice that because of Lemma 3.7 the points y_1, y_2 provide solutions to the constrained minimization problem

minimize

$$u\left(\frac{x_1+x_2}{2}\right) - u(x_1) \quad\text{such that}\quad u(x_1) \;=\; u(x_2) \quad . \tag{3.18}$$

Since $|\nabla u| \neq 0$ in D and since $u \in C^3(D)$ there exists a C^2-function $r(s)$ such that for every small real number s the relation

$$u\left(y_1 + \frac{s}{b}\,h\right) \;=\; u\left(y_2 + \frac{r(s)}{c}\,h\right) \tag{3.19}$$

holds. This follows from the implicit function theorem. We differentiate (3.19) with respect to s

$$\frac{1}{b}\,u_h\left(y_1 + \frac{s}{b}\,h\right) \;=\; \frac{r'(s)}{c}\,u_h\left(y_2 + \frac{r(s)}{c}\,h\right) \;, \tag{3.20}$$

and for $s = 0$ we obtain

$$r'(0) \;=\; 1 \quad . \tag{3.21}$$

Another differentiation of (3.20) at $s = 0$ gives

$$\frac{1}{b^2} u_{hh}(y_1) = \frac{1}{c^2} u_{hh}(y_2) + \frac{r''(0)}{c} c(n\cdot h) \quad . \tag{3.22}$$

Now let us consider the auxiliary function $\tilde{Q}(s,h) = Q\left(y_1 + \frac{s}{b} h,\ y_2 + \frac{r(s)}{c} h\right)$. For every $h \in \mathbb{R}^n$ the function $\tilde{Q}$ attains a negative minimum at $s = 0$. Therefore $\frac{\partial \tilde{Q}}{\partial s}$ has to vanish and $\frac{\partial^2 \tilde{Q}}{\partial s^2} \geq 0$. Let us calculate those two quantities and observe (3.18):

$$\frac{\partial \tilde{Q}}{\partial s}(s,h) = \frac{1}{2}\left(\frac{1}{b} + \frac{r'(s)}{c}\right) u_h \left[\frac{y_1 + y_2}{2} + \frac{1}{2}\left(\frac{s}{b} + \frac{r(s)}{c}\right) h\right] - $$

$$- \frac{1}{b} u_h \left(y_1 + \frac{s}{b} h\right) \quad . \tag{3.23}$$

For $s = 0$ we obtain the first assertion of Lemma 3.10.

$$\frac{\partial^2 \tilde{Q}}{\partial s^2}(0,h) = \left[\frac{1}{2}\left(\frac{1}{b} + \frac{1}{c}\right)\right]^2 u_{hh}\left(\frac{y_1+y_2}{2}\right) + \frac{1}{2}\frac{r''(0)}{c} a(n\cdot h) - $$

$$- \frac{1}{b^2} u_{hh}(y_1) \geq 0 \quad . \tag{3.24}$$

A combination of (3.24) (3.22) and Lemma 3.10a) yields

$$\frac{1}{a^2} u_{hh}\left(\frac{y_1+y_2}{2}\right) \geq \frac{\mu}{b^2} u_{hh}(y_1) + \frac{(1-\mu)}{c^2} u_{hh}(y_2) \quad , \tag{3.25}$$

and after summation over n orthogonal unit vectors h the proof of Lemma 3.10 is complete.

Now we are in a position to complete the proof the Theorem 3.5. We recall that $\Delta u = f(u)$ in D, and Lemmata 3.7a) and 3.10b) imply

$$\frac{1}{a^2} f\left[u\left(\frac{y_1+y_2}{2}\right)\right] \geq \left[\frac{\mu}{b^2} + \frac{(1-\mu)}{c^2}\right] f(u(y_1)) \quad . \tag{3.26}$$

If we assume for the moment that f is strictly monotone, inequality (3.26) together with (3.13) yields

$$\frac{1}{a^2} f\left[u\left(\frac{y_1+y_2}{2}\right)\right] > \left[\frac{\mu}{b^2} + \frac{(1-\mu)}{c^2}\right] f\left[u\left(\frac{y_1+y_2}{2}\right)\right] \quad . \tag{3.27}$$

Notice that because of Lemma 3.10a) and the positivity of f we obtain

$$\frac{1}{a^2} = \frac{1}{[\mu b + (1-\mu) c]^2} > \frac{\mu}{b^2} + \frac{(1-\mu)}{c^2} \quad , \tag{3.28}$$

which contradicts the convexity of $\frac{1}{x^2}$. If the function f is not strictly monotone in $(0,1)$ we can approximate it by a strictly monotone one $f_\varepsilon(u) := f(u) + \varepsilon u^+$, where $\varepsilon > 0$ is small. Then problem (3.10) has an increasing sequence of quasiconcave approximating solutions u_ε. Since $f_\varepsilon \to f$ in $L^\infty(\mathbb{R}^n)$, $u_\varepsilon \to u$ in $W^{2,\infty}(\mathbb{R}^n)$ and in particular $u_\varepsilon \to u$ uniformly on D. Therefore we can pass in (3.29) to the limit $\varepsilon \to 0$

$$Q_\varepsilon(x_1,x_2) = u_\varepsilon\left(\frac{x_1+x_2}{2}\right) - \min\left\{u_\varepsilon(x_1),\ u_\varepsilon(x_2)\right\} \geq 0 \text{ on } \mathbb{R}^n \times \mathbb{R}^n . \tag{3.29}$$

Another proof of this last limiting argument was given in [122]. This completes the proof of Theorem 3.5.

Remark 3.3

The Gabriel-Lewis method has been developed by Gabriel to prove quasiconcavity of Green's function on convex domains, and by J.L. Lewis and the author to prove quasiconcavity of capacitary functions in convex rings. Another application of this method to the linear heat equation has been given by C. Borell [35] and it appears that one can apply it to semilinear parabolic equations as well. The method can be extended to quasilinear equations of type $\operatorname{div}(|\nabla u|^{p-2}\nabla u) = f(u)$ [96].

Open problems

Under which assumptions on f are the solutions of problem (3.1) quasiconcave?

Is it true that for concave obstacle ψ the solution u of the obstacle problem (Example 2.9) in a convex domain is quasiconcave? D. Kinderlehrer and J. Spruck as well as the author have tried to prove this without success [106].

III.12 Concavity or convexity of functions

In this paragraph we shall pursue Strategy 1 from § III.11 and attempt to prove the concavity of solutions u to problem (3.1), or at least the concavity of suitable functions $g(u)$, where g is monotone increasing in its argument. An easy calculation shows that the following hierarchy of concavity holds:

$$u \text{ concave} \Rightarrow u^{\alpha} \text{ concave } [\alpha \in (0,1)] \Rightarrow \log u \text{ concave} \Rightarrow u \text{ quasiconcave} \quad .$$

The first three of these properties can be verified by proving concavity of a function $g(u)$ in the following ways:

Strategy 1a)

Show that the set $\{(x,\lambda) \in \mathbb{R}^{n+1} \mid x \in \Omega \, , \, \lambda < g(u(x))\}$ coincides with its convex hull. This was done by A. Kennington in [104, Appendix] via a maximum principle

Strategy 1b)

The parabolic approach. It was shown by Brascamp and Lieb [37] that the linear parabolic operator $\frac{\partial}{\partial t} - \Delta$ under homogeneous Dirichlet boundary conditions not only preserves positivity of initial data u_0, but also log concavity. The same is true for some semilinear parabolic operators [120]. If one denotes the solution of the corresponding initial boundary value problem by $u(x,t)$ and studies its asymptotic behavior as $t \to \infty$, then one can draw conclusions about solutions of semilinear elliptic problems and show that those have to be log concave. A function $v(x)$ is called log concave if $\log v$ is concave.

Strategy 1c)

Show that the second derivatives of $g(u)$ have nonpositive sign. This has been tried without success by N. Korevaar and the author.

Strategy 1d)

Verify inequality (3.30) below via a maximum principle. This approach is due to N. Korevaar and extensions have been found independently by L. Caffarelli, J. Spruck, A. Kennington, and the author. We shall present this strategy in detail below.

Strategy 1e)

Show that $g(u)$ can be homotopically connected to a function which is known to be convex. This continuation method was developped by L. Caferlli and A. Friedman [45].

Remark 3.4

Wa want to point out that for nonnegative $f(u)$ one cannot in general expect the solution of problem (3.1) to be concave. This can be most easily seen in the case of a plane convex domain Ω whose boundary $\partial\Omega$ is of class $C^{2,\alpha}$, $0 < \alpha < 1$, and has positive curvature, and for a function f which vanishes at the origin. Then if f is of class $C^{\alpha}(\mathbb{R})$, u is of class $C^{2,\alpha}(\bar{\Omega})$, $\frac{\partial u}{\partial n} < 0$ and $\Delta u = 0$ on the boundary. If we rewrite the Laplace operator in curvilinear coordinates we realize $\frac{\partial^2 u}{\partial n^2} > 0$ on $\partial\Omega$, so u cannot be concave.

Equivalent to the concavity of $g(u)$ is the convexity of $-g(u)$. In order to stay notationally close to N. Korevaar's paper and to point out the differences that our proof has from his, we shall check whether a function v $(= -g(u))$ is convex.

Let $v : \Omega \to \mathbb{R}$ be a continuous function and $\Omega \subset \mathbb{R}^n$ a convex bounded domain. Then v is convex if and only if

$$C(v,x_1,x_2) \quad := \quad v\left(\frac{x_1+x_2}{2}\right) - \frac{1}{2}\,v(x_1) - \frac{1}{2}\,v(x_2) \quad \leq \quad 0 \quad \text{in} \quad \Omega \times \Omega \ . \tag{3.30}$$

The function C will be called concavity function of v . If $v \in C^2(\Omega)$ is <u>not convex</u> then

i) C becomes positive as (x_1,x_2) approaches $\partial(\Omega\times\Omega)$, or

ii) C has a local positive maximum in $\Omega \times \Omega$.

The first possibility will be excluded in "boundary point lemma", Lemmata 3.11, 3.12 and the second by a "concavity maximum principle".

Lemma 3.11

Let $\Omega \subset \mathbb{R}^n$ be bounded and strictly convex, i.e. $x_1, x_2 \in \partial\Omega$ and $x_1 \neq x_2$ implies $\frac{x_1 + x_2}{2} \in \Omega$.

Let $v \in C(\Omega)$ be bounded below on $\bar{\Omega}$ and $v(x) \to +\infty$ uniformly as $d(x, \partial\Omega) \to 0$. Then case i) cannot occur, i.e. $C(v, x_1, x_2)$ cannot become positive as (x_1, x_2) approach $\partial(\Omega \times \Omega)$.

Proof

Otherwise there exist sequences $\{x_{1j}\}_{j\in\mathbb{N}}, \{x_{2j}\}_{j\in\mathbb{N}} \subset \Omega$ with limits x_1 and x_2 as $j \to \infty$ such that

$$\overline{\lim}\; C(v, x_{1j}, x_{2j}) > 0 \tag{3.31}$$

holds.

We observe that

$\frac{x_1 + x_2}{2} \in \Omega$, so that $v\left(\frac{x_1+x_2}{2}\right)$ is bounded and obtain

$$\overline{\lim}\; C(v, x_{1j}, x_{2j}) = v\left(\frac{x_1+x_2}{2}\right) - \frac{1}{2}\,\underline{\lim}\left(v(x_{1j}), v(x_{2j})\right) < 0\,, \tag{3.32}$$

a contradiction to (3.31).

Lemma 3.12

Let $\Omega \subset \mathbb{R}^n$ be bounded and strictly convex with boundary of class C^1.
Let $u \in C^1(\bar{\Omega})$ and $u > 0$ in Ω, $u = 0$ on $\partial\Omega$ and $\frac{\partial u}{\partial n} < 0$ on $\partial\Omega$, where n denotes the exterior normal to $\partial\Omega$.
Let $g : \mathbb{R}^+ \to \mathbb{R}$ be a C^1-function satisfying

$$g' > 0 \quad \text{and} \quad \lim_{u\to 0^+} g'(u) = +\infty\,. \tag{3.33}$$

Furthermore suppose that case ii) cannot occur for the concavity function of $v = -g(u)$. Then case i) cannot occur, i.e. $C(-g(u), x_1, x_2)$ cannot become positive as (x_1, x_2) approach $\partial(\Omega \times \Omega)$.

Remark 3.5

Notice that N. Korevaar and A. Kennington proved this under more severe restrictions: $\partial\Omega$ had to be smooth and have positive principal curvatures and g had to satisfy $g'' < 0$, $\lim_{u\to 0^+} \frac{g'(u)}{g''(u)} = 0$ and $\lim_{u\to 0^+} \frac{g(u)}{g'(u)} = 0$ in addition to (3.33).

If Ω satisfies the interior sphere condition, and $u \in C^2(\Omega)$ satisfies $\Delta u \leq 0$, $u > 0$ in Ω , $u = 0$ on $\partial\Omega$, then Hopf's maximum principle implies $\frac{\partial u}{\partial n} < 0$ on $\partial\Omega$.

To prove Lemma 3.12 we distinguish two cases.

Case 1

If $\lim_{u\to 0+} -g(u) = +\infty$, then we can apply Lemma 3.11, since there exist two positive constants c_1 and c_2 such that

$$c_1\, d(x,\partial\Omega) < u(x) < c_2\, d(x,\partial\Omega) \quad \text{near } \partial\Omega \ .$$

Case 2

If $\lim_{u\to 0+} g(u) = a \in \mathbb{R}$, then $v = -g(u) \in C(\overline{\Omega})$ and the concavity function $C(v,x_1,x_2) \in C(\overline{\Omega}\times\overline{\Omega})$. As a continuous function on a compact set C has to attain its maximum either in $\Omega \times \Omega$ or on the boundary. The occurance of a positive maximum in $\Omega \times \Omega$ is excluded by assymption. So let C attain a positive maximum in $(x_1,x_2) \in \partial(\Omega\times\Omega)$. If $x_1 \in \partial\Omega$ and $x_2 \in \partial\Omega$, then $C(v,x_1,x_2) < 0$ due to the positivity of u and (3.33), a contradiction. It remains to treat the possibilities $x_1 \in \partial\Omega$, $x_2 \in \Omega$ and $x_1 \in \Omega$, $x_2 \in \partial\Omega$, and without loss of generality it suffices to treat the last one. Let $n = n(x_2)$ denote the exterior normal to $\partial\Omega$ in x_2 . Then we know from the maximality of C in (x_1,x_2) :

$$d(\varepsilon) := C(v,x_1,x_2) - C(v,x_1,x_2-\varepsilon n) \geq 0 \quad \text{for } \varepsilon \geq 0 \ .$$

For positive ε the function $d(\varepsilon)$ is differentiable and $d(0) = 0$, so that

$$\underline{\lim}_{\varepsilon\to 0^+} d'(\varepsilon) \geq 0 \ . \tag{3.34}$$

If we rewrite (3.34) in terms of C we obtain

$$\lim_{\varepsilon\to 0+}\left\{\frac{1}{2}\frac{\partial v}{\partial n}\left(\frac{x_1+x_2}{2}-\frac{1}{2}\varepsilon n\right)-\frac{\partial v}{\partial n}(x_2-\varepsilon n)\right\}\ \geq\ 0\ , \tag{3.35}$$

or, since v is differentiable in $\dfrac{x_1+x_2}{2}$,

$$\frac{1}{2}g'\left[u\left(\frac{x_1+x_2}{2}\right)\right]\frac{\partial u}{\partial n}\left(\frac{x_1+x_2}{2}\right)\ \leq\ \lim_{\varepsilon\to 0+} g'\left(u(x_2-\varepsilon n)\right)\cdot\frac{\partial u}{\partial n}(x_2-\varepsilon n)\ . \tag{3.36}$$

The left hand side of (3.36) is bounded, the right hand side is unbounded from below, a contradiction. Therefore $C(-g(u),x_1,x_2)$ cannot attain a positive maximum in $\bar{\Omega}\times\bar{\Omega}$, and the proof of Lemma 3.12 is complete.

<u>Theorem 3.13</u>

<u>Let</u> $v \in C^2(\Omega)$ <u>be a solution of the elliptic equation</u>

$$\sum_{i,j=1}^{n} a^{ij}(\nabla v)\frac{\partial^2 v}{\partial x_i\,\partial x_j}-b(x,v,\nabla v)\ =\ 0\ \text{ in }\ \Omega\ . \tag{3.37}$$

<u>Let</u> $\Omega\subset\mathbb{R}^n$ <u>be bounded and convex, and let</u> $b:\Omega\times\mathbb{R}^{1+n}$ <u>satisfy</u>

$$b\geq 0\ \text{ in }\ \Omega\times\mathbb{R}^{1+n}\ ,\ b(x,u,p)-b(x,v,p)>0\ \text{ for }\ u>v\ \text{ in }\ \Omega\times\mathbb{R}^n\ , \tag{3.38}$$

<u>and</u>

b <u>is harmonic concave in</u> (x,v) (<u>i.e.</u> $\frac{1}{b}$ <u>is convex</u>):

$$\frac{1}{2}\left(b(x_1,v_1,p)+b(x_2,v_2,p)\right)b\left(\frac{x_1+x_2}{2},\frac{v_1+v_2}{2},p\right)\geq b(x_1,v_1,p)\,b(x_2,v_2,p)$$

<u>for every</u> $x_1,x_2\in\Omega$, $v_1,v_2\in\mathbb{R}$, $p\in\mathbb{R}^n$. (3.39)

<u>Then</u> $C(v,x_1,x_2)$ <u>cannot attain a positive maximum in</u> $\Omega\times\Omega$.

<u>Remark 3.6</u>

In this form, Theorem 3.13 is due to A. Kennington. He gave a proof which differs from ours; and in the author's opinion, his proof is less transparent than ours. In previous versions of this theorem assumption (3.39) was replaced by the stronger assumption [99],

$$b \text{ is log concave in } (x,u) \text{ for every fixed } p \in \mathbb{R}^n , \tag{3.40}$$

or by the even stronger assumption [112]

$$b \text{ is concave in } (x,u) \text{ for every fixed } p \in \mathbb{R}^n . \tag{3.41}$$

For the proof of Theorem 3.13 we start as in Korevaar's paper. Suppose that C attains an interior positive maximum at a pair $(x_1,x_3) \in \Omega\times\Omega$, i.e.

$$v(x_2) > \frac{1}{2} v(x_1) + \frac{1}{2} v(x_3) , \quad \text{where } x_2 := \frac{1}{2}(x_1+x_3) . \tag{3.42}$$

Since the gradient of C with respect to its $2n$ coordinates has to vanish at (x_1,x_3) we obtain

$$\nabla v(x_1) = \nabla v(x_2) = \nabla v(x_3) . \tag{3.43}$$

Now consider the restricted concavity function $\tilde{C}$, which is defined in a neighborhood of (x_1,x_3) by translating x_1 by αw and x_3 by βw, where w is a small vector in $\mathbb{R}^n$ and α,β are real numbers such that the arguments in (3.44) below lie in Ω.

$$\tilde{C}(w,\alpha,\beta) := v\left(x_2 + \frac{\alpha+\beta}{2} w\right) - \frac{1}{2} v(x_1+\alpha w) - \frac{1}{2} v(x_2+\beta w) . \tag{3.44}$$

This is the point where our proof differs from Korevaar's and Kennington's. Korevaar had introduced $\bar{C}(w) := u(x_2+w) - \frac{1}{2} u(x_1+w) - \frac{1}{2}(x_3+w)$ and Kennington had essentially introduced

$$\bar{C}(y_1,y_3) := u\left(x_2 + \frac{y_1+y_3}{2}\right) - \frac{1}{2} u(x_1+y_1) - \frac{1}{2} u(x_3+y_3) .$$

For each (α,β) in $\mathbb{R}^2$ the function $\tilde{C}(w,\alpha,\beta)$ has a local maximum at $w = 0$. Hence

$$D_w \tilde{C}(0,\alpha,\beta) = 0 \quad \text{and} \quad [D_w^2 \tilde{C}(0,\alpha,\beta)] \leq 0 , \tag{3.45}$$

where the symbols in (3.45) represent the gradient and the Hessean matrix of $\tilde{C}$ with respect to w. Now let a^{ij} and $b(x,v)$ be shorthand for $a^{ij}(\nabla v)$ and $b(x,v,\nabla v)$ at the common values (3.43) of ∇v. Now (3.45) and the ellipticity of the matrix a^{ij} imply

$$\sum_{i,j=1}^{n} a^{ij} \, [D_w^2 \, \tilde{C}(0,\alpha,\beta)]_{i,j} \leq 0 \quad ,$$

i.e. using the summation convention and the notation $v_{ij} = \frac{\partial^2 v}{\partial x_i \, \partial x_j}$ we obtain

$$a^{ij} \left[\left(\frac{\alpha+\beta}{2}\right)^2 v_{ij}(x_2) - \frac{\alpha^2}{2} v_{ij}(x_1) - \frac{\beta^2}{2} v_{ij}(x_3) \right] \leq 0 \quad . \tag{3.46}$$

Using (3.46) and the differential equation (3.37) we arrive at the crucial inequality

$$\left(\frac{\alpha+\beta}{2}\right)^2 b(x_2,v(x_2)) \leq \frac{\alpha^2}{2} b(x_1,v(x_1)) + \frac{\beta^2}{2} b(x_3,v(x_3)) \quad . \tag{3.47}$$

Notice that (3.47) represents a whole family of inequalities, since α and β can be freely chosen. For $\alpha = \beta = 1$ this reduces to Korevaar's key inequality

$$b(x_2,v(x_2)) \leq \frac{1}{2} b(x_1,v(x_1)) + \frac{1}{2} b(x_3,v(x_3)) \quad , \tag{3.48}$$

and under Korevaar's concavity assumption (3.41) one can easily derive a contradiction. We want to use the weaker assumption (3.39) and have to make other choices of α and β. First we note that (3.42) and (3.38) imply

$$b(x_2,v(x_2)) > b\left(x_2, \frac{1}{2}(v(x_1)+v(x_3))\right) \geq 0 \quad , \tag{3.49}$$

so that a choice of $\alpha = 1$, $\beta = 0$ in (3.47) yields, using (3.49)

$$b(x_1,v(x_1)) \geq \frac{1}{2} b(x_2,v(x_2)) > 0 \quad . \tag{3.50}$$

Now we choose $\alpha = b(x_3,v(x_3))$ and $\beta = b(x_1,v(x_1))$. Then (3.47) reduces to

$$\left(\frac{\alpha+\beta}{2}\right)^2 b(x_2,v(x_2)) \leq \alpha \, \beta \left(\frac{\alpha+\beta}{2}\right) \quad . \tag{3.51}$$

We observe that (3.50) and (3.38) imply $\frac{\alpha+\beta}{2} > 0$ and divide (3.51) by $\frac{\alpha+\beta}{2}$. This and (3.49) lead to

$$\frac{\alpha + \beta}{2} \, b \left(x_2, \frac{1}{2} (v(x_1)+v(x_3)) \right) < \alpha \beta \, , \qquad (3.52)$$

which contradicts (3.39).

Remark 3.7

The proof of Theorem 3.13 contains some particular choices of α and β. One might wonder whether different choices of α and β would allow for a weaker assumption on b than (3.39). Let us dispel any hopes about this. Our proof works as long as (3.47) leads to a contradiction. Inequality (3.47) can be interpreted as stating that a certain polynomial in α and β attains values of prescribed sign only. If we look for nontrivial zeros of this polynomial, condition (3.39) shows up as a necessary and sufficient condition. This interpretation of inequality (3.47) and a more complicated derivation of (3.47) are the essential ideas in A. Kenningtons proof. Nevertheless his proof reveals the optimality of (3.39).

Remark 3.8

One can easily state and prove analogous theorems for parabolic equations. The interested reader may use. N. Korevaar's paper as a guideline.

Again we want to give some examples.

Example 3.3 Conformal incenter, an open problem of Caffarelli and Friedman.

This problem was suggested to the author by C. Bandle and G. Keady and can also be found in A. Friedman's book [75, p. 535, 547 f.] and in the paper [43] of L. Caffarelli and A. Friedman. Let $\Omega \subset \mathbb{R}^2$ be a strictly convex bounded domain with boundary $\partial\Omega$ of class $C^{2,\gamma}$.

Consider the boundary value problem

$$\Delta u = b\, e^{du} \quad \text{in } \Omega \, , \qquad (3.53)$$

$$u \to +\infty \text{ uniformly as } d(x,\partial\Omega) \to 0 \, , \qquad (3.54)$$

where b,d are positive constants. Equation (3.53) is also known as Liouville's equation [19, p. 67; 121]. Of mathematical interest is the conformal incenter.

$$S := \left\{x \in \Omega \,|\, u \text{ attains its minimum in } x\right\} .$$

In [75] it is stated that S is a finite set and in [43] Caffarelli and Friedman ask for conditions under which S is singleton.

Theorem 3.13 and Lemma 3.12 imply not only that S is <u>singleton</u> for strictly convex domains Ω with boundary of class $C^{2,\gamma}$, but also that the solution u to (3.53) (3.54) is convex.

<u>Remark 3.9</u>

Example 3.3 is related to the plasma problem, Example 2.14, in the following way. For $\lambda \to \infty$ and for variational solutions u of the plasma problem, the plasma region Ω_o "shrinks" to a point in S. Formally, if we denote solutions associated to λ by u_λ, we expect $\lim_{\lambda\to\infty} u_\lambda^+ = 0$ to hold. In terms of singular perturbation theory we expect an interior boundary layer near S. This was in fact shown by L. Caffarelli and A. Friedman [43, 172].

<u>Example 3.4</u> Saint Venant torsion problem.

Let $\Omega \subset \mathbb{R}^n$ be a strictly convex domain and let u be the solution of the Saint Venant torsion problem

$$\Delta u = -1 \quad \text{in } \Omega , \tag{3.55}$$

$$u = 0 \quad \text{on } \partial\Omega . \tag{3.56}$$

The function u is also known as warping function. For $n = 2$ it is known that the square root of u is a strictly concave function [95]. For $n \geq 2$ the concavity of $u^{1/2}$ follows from Lemma 3.12 and Theorem 3.13. If we set $v = h(u) = -\sqrt{u}$, then v solves the boundary value problem

$$\Delta v = -\frac{1}{v}\left(|\nabla v|^2 + \frac{1}{2}\right), \quad v < 0 \text{ in } \Omega \tag{3.57}$$

$$v = 0 \quad \text{on } \partial\Omega . \tag{3.58}$$

Consequently v has to be convex, i.e. <u>$u^{1/2}$ is concave</u>.

Example 3.5 Nonlinear diffusion.

This problem was suggested to me by N. Alikakos. Consider the following initial boundary value problem for $m > 1$:

$$\frac{\partial w}{\partial t}(x,t) - \Delta(w^m(x,t)) = 0 \quad \text{in} \quad \Omega \times [0,\infty) \ , \tag{3.59}$$

$$w(x,t) = 0 \quad \text{on} \quad \partial\Omega \times [0,\infty) \ , \tag{3.60}$$

$$w(x,0) = w_o(x) \geq 0 \quad \text{in} \quad \Omega \ , \tag{3.61}$$

which describes flow through a porous medium. D. Aronson and L.A. Peletier showed that w^m tends assymptotically for $t \to \infty$ to a separable solution, whose spatial factor $u(x)$ is the (unique) positive solution of the following problem

$$\Delta u(x) + \frac{1}{m-1} u^{1/m}(x) = 0 \quad \text{in} \quad \Omega \ ,$$

$$u(x) = 0 \quad \text{on} \quad \partial\Omega \ .$$

In this context [12] the function u is called the asymptotic profile of w . Thus we are led to study positive solutions of

$$\Delta u + \lambda u^q = 0 \ , \quad u > 0 \quad \text{in} \quad \Omega \ , \tag{3.62}$$

$$u = 0 \quad \text{on} \quad \partial\Omega \ , \tag{3.63}$$

where $\Omega \subset \mathbb{R}^n$ is a strictly convex bounded domain with boundary $\partial\Omega$ of class $C^{2,\gamma}$, $\lambda > 0$ and $0 < q < 1$. Then the substitution $v = -u^{(1-q)/2}$ leads to the differential equation

$$\Delta v = -\frac{1}{v}\left(\frac{1+q}{1-q}|\nabla v|^2 + \lambda(1-q)/2\right) \ . \tag{3.64}$$

By the same reasoning as in the previous example we obtain that <u>$u^{(1-q)/2}$ has to be concave</u>.

Open problem

Prove uniqueness of solutions to (3.62) (3.63) for $\lambda > 0$, $1 < q < \frac{n+2}{n-2}$.

If Ω is a ball, there is uniqueness [85]; if Ω is an annulus, there is no uniqueness [38]. A geometric assumption on Ω which induces uniqueness might be convexity. This problem plays a role in "fast diffusion" [34, 144].

Remark 3.10

If u solves $\Delta u + f(u) = 0$, $u > 0$ in Ω , $u > 0$ in Ω , $u = 0$ on $\partial\Omega$, and if $\Omega \subset \mathbb{R}^n$ is convex, then

$$g(u(x)) := \int^{u(x)} [F(s)]^{-1/2} \, ds \quad \text{with} \quad F(t) := \int_0^t f(s) \, ds$$

is a good candidate for being concave. This is the case for Examples 3.4 and 3.5, and more on this question can be found in [101].

REFERENCES

[1] Acker, A. On the convexity of equilibrium plasma configurations. Math. Methods Appl. Sci. 3 (1981) 435-443

[2] Acker, A., L.E. Payne, G. Philippin On the convexity of level lines of the fundamental mode in the clamped membrane problem, and the existence of convex solutions in a related free boundary problem. Z. Angew. Math. Phys. 32 (1981) 683-694

[3] Adams, R.A. Sobolev spaces, Acad. Press, New York 1975

[4] Alt, J.W., L.A. Caffarelli Existence and regularity for a minimum problem with free boundary, J. Reine Angew. Math. 325 (1981) 105-114

[5] Alt, H.W., L.A. Caffarelli, A. Friedman A free boundary problem for quasilinear elliptic equations. Ann. Sc. Norm. Sup., Pisa, IV. Ser., 11 (1984) 1-44

[6] Alt, H.W., D. Phillips A free boundary problem for semilinear elliptic equations, preprint 709 SFB 72 Bonn 1985

[7] Alvino, A., P.L. Lions, G. Trombetti A remark on comparison results via symmetrization, manuscript 1985

[8] Alvino, A., G. Trombetti Sulle migliori costanti di maggiorazione per una classe di equazioni ellittiche degeneri. Ric. Mat. 27 (1978) 413-428

[9] Alvino, A., G. Trombetti Equazione ellitiche con termine di ordine inferiore e riordinamenti. Atti Accad. Naz. Lincei, Mem., Cl. Sci. Fis. Mat. Nat., VIII. Ser., Sez I 66 (1979) 1-7

[10] Alvino, A., G. Trombetti A lower bound for the first eigenvalue of an elliptic operator. J. Math. Anal. Appl. 94 (1983) 328-337

[11] Alvino, A., G. Trombetti Isoperimetric inequalities connected with torsion problem and capacity. preprint 60, 1983 Napoli

[12] Aronson, D.G., L.A. Peletier Large time behaviour of solutions of the porous medium equation in bounded domains. J. Differ. Equations 39 (1981) 378-412

[13] Aronsson, G. An integral inequality and plastic torsion. Arch. Ration. Mech. Anal. 72 (1979) 23-39

[14] Aronsson, G., G. Talenti Estimating the integral of a function in terms of a distribution function of its gradient. Boll. Unione Mat. Ital., V. Ser., B 18 (1981) 885-894

[15] Auchmuty, G. Existence of axisymmetric equilibrium figures. Arch. Ration. Mech. Anal. 35 (1977) 249-261

[16] Ash, R.B. Real Analysis and Probability, Acad. Press, New York 1972

[17] Baiocchi, C. Su un problema a frontiera libera connesso a questioni di idraulica. Ann. Mat. Pura Appl. 92 (1972) 107-127

[18] Bailet-Intissar, J. Non linear mixed boundary value problems for elliptic partial differential equations. Proc. R. Soc. Edinb., Sec. A 90 (1981) 347-359

[19] Bandle, C. Isoperimetric Inequalities and Applications, Pitman, London 1980

[20] Bandle, C., M. Marcus Radial averaging transformations with various metrics. Pac. J. Math. 46 (1973) 337-348

[21] Bandle, C., M. Marcus Radial averaging transformations and generalized capacities. Math. Z. 145 (1975) 11-17

[22] Bandle, C., M. Marcus A priori estimates and the boundary values of solutions for a problem arising in plasma physics. Nonlinear Anal., Theory Methods Appl. 7 (1983) 439-451

[23] Bandle, C., M. Marcus On the size of the plasma region. Appl. Anal. 15 (1983) 207-225

[24] Bandle, C., J. Mossino Application du réarrangement à une inéquation variationnelle.C.R. Acad. Sci., Paris, Ser. I 296 (1983) 501-504

[25] Bandle, C., J. Mossino Rearrangements in variational inequalities. Ann. Mat. Pura Appl., IV. Ser. 138 (1984) 1-14

[26] Bandle, C., R. Sperb Qualitative behaviour and bounds in a nonlinear plasma problem. SIAM J. Math. Anal. 14 (1983) 142-151

[27] Bandle, C., R. Sperb, I. Stakgold Diffusion reaction with monotone kinetics. Nonlinear Anal., Theory Methods Appl. 8 (1984) 321-333

[28] Bandle, C., I. Stakgold Isoperimetric inequality for the effectiveness in semilinear parabolic problems. ISNM 71 (1984) 289-295

[29] Bandle, C., I. Stakgold The formation of the dead core in parabolic reaction diffusion problems. Trans. Am. Math. Soc. 286 (1984) 275-293

[30] Bardi, M. Geometric properties of solutions of Hamilton Jacobi equations. J. Differ. Equations, to appear

[31] Barnes, D.C. Rearrangement of functions and lower bounds for eigenvalues of differential equations. Appl. Anal. 13 (1982) 237-248

[32] di Benedetto, E. $C^{1+\alpha}$ local regularity of weak solutions of degenerate elliptic equations. Nonlinear Anal., Theory Methods Appl. 7 (1983) 827-850

[33] Berger, M. Nonlinearity and Functional Analysis, Acad. Press, New York 1972

[34] Berryman, J.G., C.J. Holland Stability of the separable solution for fast diffusion. Arch. Ration. Mech. Anal. 74 (1980) 379-388

[35] Borell, C. Brownian motion in a convex ring and quasiconcavity. Commun. Math. Phys. 86 (1982) 143-147

[36] Brascamp, H.J., E.H. Lieb, J.M. Luttinger A general rearrangement inequality for multiple integrals. J. Funct. Anal. 17 (1974) 227-237

[37] Brascamp, H.J., E.H. Lieb On extensions of the Brunn-Minkowski and Prékopa-Leindler theorems, including inequalities for log-concave functions, and with an application to the diffusion equation. J. Funct. Anal. 22 (1976) 366-389

[38a] Brézis, H., E. Lieb Sobolev inequalities with remainder terms. J. Funct. Anal. 62 (1985) 73-86

[38b] Brézis, H., L. Nirenberg Positive solutions of nonlinear elliptic equations involving critical Sobolev exponents. Commun. Pure Appl. Math. 36 (1983) 437-477

[39] Brézis, H. Problems unilatéraux. J. Math. Pures Appl. 51 (1972) 1-168

[40] Buononcore, P. Second order elliptic equations whose coefficients have their first derivatives in Lorentz spaces. Ric. Mat. 32 (1983) 187-202

[41] Buononcore, P. Sur la rigidité à la torsion des domaines doublement connexes. C.R. Acad. Sci., Paris, Ser. I, 298 (1984) 241-244

[42] Buononcore, P. Isoperimetric inequalities in the torsion problem for multiply connected domains. Z. Angew. Math. Phys. 36 (1985) 47-60

[43] Caffarelli, L.A., A. Friedman Asymptotic estimates for the plasma problem. Duke Math. J. 47 (1980) 705-742

[44] Caffarelli, L.A., A. Friedman Partial regularity of the zero set of solutions of linear and superlinear elliptic equations. manuscript 1984

[45] Caffarelli, L.A., A. Friedman Convexity of solutions of semilinear elliptic equations. manuscript 1984

[46] Caffarelli, L.A., J. Spruck Convexity properties of solutions to some classical variational problems. Commun. Partial Differ. Equations 7 (1982) 1337-1379

[47] Casten, R.G., C.J. Holland Instability results for reaction diffusion equations with Neumann boundary conditions. J. Differ. Equations 27 (1978) 266-273

[48] Chafee, N. Asymptotic behavior for solutions of a one-dimensional parabolic equation with homogeneous Neumann boundary conditions. J. Differ. Equations 18 (1975) 111-135

[49] Chipot, M., J.K. Hale Stable equilibria with variable diffusion. Contemporary Mathematics 17 (1983) 209-213

[50] Chiti, G. Rearrangements of functions and convergence in Orlicz spaces. Appl. Anal. 9 (1979) 23-27

[51] Chiti, G. An isoperimetric inequality for the eigenfunctions of linear second order elliptic operators. Boll. Unione Mat. Ital., VI. Ser., A 1 (1982) 145-151

[52] Chiti, G. A reverse Hölder inequality for the eigenfunctions of linear second order elliptic operators. Z. Angew. Math. Phys. 33 (1982) 143-148

[53] Chiti, G. A bound for the ratio of the first two eigenvalues of a membrane. SIAM J. Math. Anal. 14 (1983) 1163-1167

[54] Ciarlet, P.G., P.A. Raviart General Lagrange and Hermite interpolation in $\mathbb{R}^n$ with application to finite element methods. Arch. Ration. Mech. Anal. 46 (1972) 177-199

[55] Coffmann, C.V. A nonlinear boundary value problem with many possitive solutions. J. Differ. Equations 54 (1984) 429-437

[56a] Coron, J.M. La continuité de la symétrization dans l'ensemble des fonctions positives de $H^1(\mathbb{R})$. C.R. Acad. Sci., Paris, Ser. I 293 (1981) 357-360

[56b] Coron, J.M. The continuity of rearrangement in $W^{1,p}(\mathbb{R})$. Ann. Sc. Norm. Sup., Pisa, IV. Ser., 11 (1984) 57-85

[57] Courant, R., D. Hilbert Methoden der Mathematischen Physik. Springer-Verlag, Berlin 1968

[58] Crandall, M.G., L. Tartar Some relations between nonexpansive and order preserving mappings. Proc. Am. Math. Soc. 78 (1980) 385-390

[59] Crooke, P.S., L.E. Payne Continuous dependence on geometry for the backward heat equation. Math. Methods Appl. Sci. 6 (1984) 433-448

[60] Cryer, C.W. A proof of the convexity of the free boundary for porous flow through a rectangular dam using the maximum principle. J. Inst. Math. Applics. 25 (1980) 111-120

[61] Diaz, J.I., M.A. Herrero Estimates on the support of the solutions of some nonlinear elliptic and parabolic problems. Proc. R. Soc. Edinb. Sect. A 89 (1981) 347-359

[62] Duff, G.F.D. Differences, derivatives and decreasing rearrangements. Canad. J. Math. 19 (1967) 1153-1178

[63] Duff, G.F.D. Integral inequalities for equimeasurable rearrangements. Canad. J. Math. 22 (1970) 408-430

[64] Duff, G.F.D. A general integral inequality for the derivative and an equimeasurable rearrangement. Canad. J. Math. 28 (1976) 793-804

[65] Egnell, H. Extremal properties of the first eigenvalue of a class of elliptic eigenvalue problems. Report 7, Uppsala 1985

[66a] Eisen, G. On the obstacle problem with a volume constraint. manuscr. math. 43 (1983) 73-83

[66b] Eisen, G. Variational problems with obstacles and integral constraints. Appl. Math. Optimization 12 (1984) 173-189

[67] Esteban, M.J., P.L. Lions Existence and non-existence results for semilinear elliptic problems in unbounded domains. Proc. R. Soc. Edinb. Sect. A 93 (1982) 1-14

[68] Evans, L.C. A new proof of local $C^{1,\alpha}$ regularity for solutions of certain degenerate elliptic partial differential equations. J. Differ. Equations 45 (1982) 356-373

[69] Faber, G. Beweis, daß unter allen homogenen Membranen von gleicher Fläche und gleicher Spannung die kreisförmige den tiefsten Grundton gibt. Sitzungsber. Bayer. Akad. Wiss., Math.-Naturwiss. Kl. (1923) 169-172

[70] Federer, H. Curvature measure. Trans. Am. Math. Soc. 93 (1959) 418-491

[71] Fichera, G. Problemi elastostatici con vincoli unilaterali: il problema di Signorini con ambigue condizioni al contorno. Atti Accad. Naz. Lincei, Mem., VIII. Ser., 1. Sez. 7 (1964) 91-140

[72] de Figuereido, D.G., P.L. Lions, R.D. Nussbaum A priori estimates and existence of positive solutions of semilinear elliptic equations. J. Math. Pures Appl. 61 (1982) 41-63

[73] Fraenkel, L.E., M.S. Berger A global theory of steady vortex rings in an ideal fluid. Acta Math. 132 (1974) 13-51

[74] Frank, L.S., W.D. Wendt On an elliptic operator with discontinuous nonlinearity. J. Differ. Equations 54 (1984) 1-18

[75] Friedman, A. Variational principles and free boundary problems. Wiley, New York 1982

[76] Friedman, A., B. McLeod Strict inequalities for integrals of decreasingly rearranged functions. manuscript 1984

[77] Friedman, A., D. Phillips The free boundary of a semilinear elliptic equation. Trans. Am. Math. Soc. 282 (1984) 153-182

[78] Gabriel, R. A result concerning convex level surfaces of 3-dimensional harmonic functions. J. London Math. Soc. 32 (1957) 286-294

[79] Gallouet, T. Contribution à l'étude d'une équation apparaissant en physique des plasmas. Thesis, Paris 1978

[80] Garabedian, P.R., H. Lewy, M. Schiffer Axially symmetric cavitational flow. Ann. Math. 56 (1952) 560-602

[81] Gehring, F.W. Symmetrization of rings in space. Trans. Am. Math. Soc. 101 (1961) 499-519

[82] Giarusso, E. Un'osservazione su un equazione parabolica del II ordine. Rend. Accad. Sci. Fis. Mat., IV. Ser., Napoli 48 (1980/81) 161-170

[83] Giarusso, E. Su una classe di equazione non lineari ellittiche. Ric. Mat. 31 (1982) 245-257

[84] Giarusso, E., D. Nunziante Symmetrization in a class of first order Hamilton-Jacobi equations. Nonlinear Anal., Theory Methods Appl. 8 (1984) 289-299

[85] Gidas, B., W.M. Ni, L. Nirenberg Symmetry and related properties via the maximum principle. Comm. Math. Phys. 68 (1979) 209-243

[86] Grabmüller, H. A note on equimeasurable starshaped rearrangement or: "How long is John Miller's nose?". Report 096, Erlangen (1983)

[87] Grabmüller, H. private communication

[88] Hardy, G.H., J.E. Littlewood, G. Polya Inequalities. Cambridge Univ. Press, Cambridge 1934

[89] Hayman, W.K. Multivalent functions. Cambridge Univ. Press, Cambridge 1958

[90] Hersch, J. private communication

[91] Hilden, Keijo Symmetrization of functions in Sobolev spaces and the isoperimetric inequality. manuscr. math. 18 (1976) 215-235

[92] Jones, C.K.R.T. Asymptotic behaviour of a reaction diffusion equation in higher space dimension. Rocky Mt. J. Math. 13 (1983) 355-364

[93] Kawohl, B. Starshapedness of level sets for the obstacle problem and for the capacitory potential problem. Proc. Am. Math. Soc. 89 (1983) 637-640

[94] Kawohl, B. A geometric property of level sets of solutions to semilinear elliptic Dirichlet problems. Appl. Anal. 16 (1983) 229-234

[95] Kawohl, B. When are superharmonic functions concave? Applications to the St. Venant torsion problem and to the fundamental mode of the clamped membrane. Z. Angew. Math. Mech. 64 (1984) 364-366

[96] Kawohl, B. Geometrical properties of level sets of solutions to elliptic problems. Proc. AMS-Summer Research Institute on Nonlinear Functional Analysis and Applications in Berkeley 1983. Ed. F. Browder. to appear in Proc. of Symposia in Pure Math. 44 (1985)

[97] Kawohl, B. Starshaped rearrangement and applications. LCDS Report 83-20, submitted

[98] Kawohl, B. On the isoperimetric nature of a rearrangement inequality and its consequences for some variational problems. LCDS Report 84-4, Arch. Ration. Mech. Anal., submitted

[99] Kawohl, B. A remark on N. Korevaar's concavity maximum principle and on the asymptotic uniqueness of solutions to the plasma problem. LCDS Report 84-6, Math. Methods Appl. Sci., to appear

[100] Kawohl, B. On the convexity and symmetry of solutions to an elliptic free boundary problem. LCDS Report 84-10, submitted

[101] Kawohl, B. When are solutions to nonlinear elliptic boundary value problems convex? Commun. Partial Differ. Equations, to appear

[102] Keady, G. An elliptic boundary value problem with a discontinuous nonlinearity. Proc. R. Soc. Edinb., Sect. A 91 (1981) 161-174

[103] Keller, H.B., J.K. Keller Lowest eigenvalue of nearly circular regions. Q. Appl. Math. 12 (1954) 141-150

[104] Kennington, A. An improved convexity maximum principle and some applications. Thesis, Adelaide, Feb. 1984

[105a] Kiener, K. Extremalität von Ellipsoiden und die Faltungsgleichung von Riesz-Sobolev. manuscript 1984

[105b] Kinderlehrer, D. Variational inequalities and free boundary problems. Bull. Am. Math. Soc. 84 (1978) 7-26

[106] Kinderlehrer, D., J. Spruck private communication

[107] Kinderlehrer, D., G. Stampacchia An introduction to variational inequalities and their applications. Acad. Press, New York 1980

[108] Kishimoto, K., H.F. Weinberger The spatial homogeneity of stable equilibria of some reaction diffusion systems on convex domains. preprint 1984

[109] Klemes, I. A mean oscillation inequality. Proc. Am. Math. Soc. 93 (1985) 497-500

[110] Kohler-Jobin, M.T. Démonstration de l'inégalité isopérimétrique $P \lambda^2 \geq \pi j_0^4/2$ conjecturée par Pólya et Szegö. C.R. Acad. Sci., Paris, Ser. I 281 (1975) 119-121

[111] Kohler-Jobin, M.T. Symmetrization with equal Dirichlet integrals. SIAM J. Math. Anal. 13 (1982) 153-161

[112] Korevaar, N. Convex solutions to nonlinear elliptic and parabolic boundary value problems. Indiana Univ. Math. J. 32 (1983) 603-614

[113] Krahn, E. Über eine von Rayleigh formulierte Minimaleigenschaft des Kreises. Math. Ann. 94 (1924) 97-100

[114] Leichtweiß, K. Konvexe Mengen. Springer-Verlag, Berlin 1980

[115] Lewis, J.L. Capacitary functions in convex rings. Arch. Ration. Mech. Anal. 66 (1977) 201-224

[116] Lieb, E.H. Existence and uniqueness of the minimizing solution of Choquard's non linear equation. Studies in Appl. Math. 57 (1977) 93-105

[117] Lieb. E. Sharp constants in the Hardy-Littlewood and related inequalities. Ann. Math. 118 (1983) 349-374

[118] Lions, P.L. Quelques remarques sur la Symétrization de Schwartz, Pitman Res. Notes in Math. 53 (1981) 308-319

[119] Lions, P.L. On the existence of positive solutions of semilinear elliptic equations. SIAM Rev. 24 (1982) 441-467

[120] Lions, P.L. Two geometrical properties of solutions of semilinear problems. Appl. Anal. 12 (1981) 267-272

[121] Liouville, J. Sur l'équation aux différences partielles $\frac{d^2 \log\lambda}{du\, dv} \pm \frac{\lambda}{2a^2}$. J. Math. Pures Appl. 18 (1853) 71-72

[122] Longinetti, M. Compattezza di successioni di funzioni quasi convesse equilimitate. Boll. Unione Mat. Ital., V. Ser. A 17 (1980) 338-344

[123] Longinetti, M. An inequality for quasiconvex functions. Appl. Anal. 13 (1982) 93-96

[124] Longinetti, M. Sulla convessita delle linee di livello di funzione armoniche. Boll. Unione Mat. Ital., VI. Ser. A 2 (1983) 71-76

[125] Longinetti, M. A maximum principle for the starshape of solutions of nonlinear Poisson equations. Boll. Unione Mat. Ital., VI. Ser. A 3 (1984)

[126] Luttinger, J.M. Generalized isoperimetric inequalities I, II and III. J. Math. Phys. 14 (1973), 586-593, 1444-1447 and 1448-1450

[127] Luttinger, J.M. Generalized isoperimetric inequalities. Proc. Nat. Acad. Sci. USA 70 (1973) 1005-1006

[128] Maderna, C., S. Salsa Symmetrization in Neumann problems. Appl. Anal. 9 (1979) 247-256

[129] Makar-Limanov, L.G. Solution of Dirichlet's problem for the equation $\Delta u = -1$ on a convex region. Math. Notes Acad. Sci. USSR 9 (1971) 52-53

[130] Marcus, M. Transformation of a domain in the plane and applications in the theory of functions. Pac. J. Math. 14 (1964) 613-626

[131] Marcus, M. Radial averaging of domains, estimates for Dirichlet integrals and applications. J. Anal. Math. 27 (1974) 47-93

[132] Marti, J.T. The least constant in Friedrichs' inequality in one dimension. SIAM J. Math. Anal. 16 (1985) 148-150

[133] Mascolo, E., L. Migliaccio, R. Schianchi An inequality for L^∞-norm of eigenfunctions of linear second order elliptic operators. Boll. Unione Mat. Ital., VI. Ser. C 1 (1982) 51-60

[134] Matano, H. Nonincrease of the lap number of a solution for a one-dimensional semilinear parabolic equation. J. Fac. Sci., Univ. Tokyo, Sect. I A 29 (1982) 401-441

[135] Matano, H. Asymptotic behaviour and stability of solution of semilinear diffusion equations. Publ. Res. Inst. Math. Sci. 15 (1979) 401-454

[136] Matano, H. Asymptotic behaviour of the free boundaries arising in one phase Stefan problems in multidimensional spaces. Lecture Notes in Num. Appl. Anal. 5 (1982) 133-151

[137] Meyer, G.H. Free boundary problems with nonlinear source terms. Numer. Math. 43 (1984) 463-482

[138] Mitrinović, D.S. Analytic Inequalities. Springer-Verlag, Berlin 1980

[139] Morrey, C.B., L. Nirenberg On the analyticity of the solutions of linear elliptic systems of partial differential equations. Commun. Pure Appl. Math. 10 (1957) 271-290

[140] Mossino, J. Inégalités isopérimétriques et applications en physique. Herman, Paris 1984

[141] Muheim, J.A. Verfahren zur Berechnung der akustischen Eigenfrequenzen und Stehwellenfelder komplizierter Hohlräume. Dissertation, ETH Zürich No. 4810, 1972

[142] Nečas, J. Les méthodes directes en théorie des équations elliptiques. Academia, Prague 1967

[143] Ni, W.M., P. Sacks The number of peaks of positive solutions of semilinear parabolic equations. SIAM J. Math. Anal., to appear

[144] Ni. W.M., P. Sacks, J. Tavantzis On the asymptotic behavior of solutions of certain quasilinear parabolic equations. J. Differ. Equations 54 (1984) 97-120

[145] Nickel, K. Gestaltaussagen über Lösungen parabolischer Differentialgleichungen. J. Reine Angew. Math. 211 (1962) 78-94

[146] Nitsche, J.C.C. Variational problems with inequalities as boundary conditions or how to fashion a cheap hat for Giacometti's brother. Arch. Ration. Mech. Anal. 35 (1969) 83-113

[147] Novak, E. Two remarks on the decreasing rearrangement of a function. Report 108, Erlangen 1984

[148] Payne, L.E. Isoperimetric inequalities and their applications. SIAM Rev. 9 (1967) 453-488

[149] Payne, L.E. On two conjectures in the fixed membrane eigenvalue problem. Z. Angew. Math. Phys. 24 (1973) 721-729

[150] Payne, L.E., G.A. Phillipin Isoperimetric inequalities in the torsion and clamped membrane problem for convex plane domains. SIAM J. Math. Anal. 14 (1983) 1154-1162

[151] Payne, L.E., A. Weinstein Capacity, virtual mass and generalized symmetrization. Pac. J. Math. 2 (1952) 633-641

[152] Pfaltzgraff, J.A. Radial symmetrization and capacities in space. Duke Math. J. 34 (1967) 747-756

[153] Pohožaev, S.I. Eigenfunctions of the equation $\Delta u + \lambda f(u) = 0$. Soviet Math. Dokl. 165 (1965) 1408-1411

[154] Polster, A. private communication

[155] Polya, G. Circle, sphere, symmetrization, and some classical physical problems. in: Modern mathematics for the engineer, ed. E.F. Beckenbach, 2nd series. McGraw-Hill, New York 1961

[156] Polya, G., G. Szegö Isoperimetric inequalities in mathematical physics. Ann. Math. Stud. 27 (1952), Princeton Univ. Press

[157] Protter, M.H., H.F. Weinberger Maximum principles in differential equations. Prentice Hall Inc., Englewood Cliffs N.J. 1967

[158] Riesz, F. Sur une inégalité intégrale. J. Lond. Math. Soc. 5 (1930) 162-168

[159] Royden, H.L. Real Analysis. McMillan, New York 1968

[160] Ryff, J.K. Measure preserving transformations and rearrangements. J. Math. Anal. Appl. 31 (1970) 449-458

[161] Sakaguchi, S. Starshaped coincidence sets in the obstacle problem. Ann. Sc. Norm. Sup., Pisa, IV. Ser., 11 (1984) 123-128

[162] Schaeffer, D.G. An example of the plasma with infinitely many solutions. unpublished manuscript 1976

[163] Schaeffer, D.G. Nonuniqueness in the equilibrium shape of a confined plasma. Commun. Partial Differ. Equations 2 (1977) 587-600

[164] Schaible, S., W.T. Ziemba Generalized concavity in optimization and economics. Proc. of a NATO Conference in Vancouver 1980, Acad. Press, New York 1981

[165] Schwarz, B. Bounds for the principal frequency of a nonhomogeneous membrane and for the generalized Dirichlet integral. Pac. J. Math. 7 (1957) 1653-1676

[166] Schwarz, B. On the extrema of the frequencies of nonhomogeneous strings with equimeasurable density. J. Math. Mech. 10 (1961) 401-422

[167] Schwarz, H.A. Gesammelte Mathematische Abhandlungen. Springer-Verlag, Berlin 1890

[168] Schwarz, H.R. Methode der finiten Elemente. Teubner, Stuttgart 1980

[169] Serrin, J. A symmetry problem in potential theory. Arch. Ration. Mech. Anal. 43 (1971) 304-318

[170] Simader, C.G. Remarks on uniqueness and stability of weak solutions of strongly nonlinear elliptic equations. Bayreuther Math. Schriften 11 (1982) 67-79

[171] Simon, J. Regularité de la solution d'un problème aux limites non linéaires. Ann. Fac. Sci. Toulouse, V. Ser., Math. 3 (1981) 247-274

[172] Simon, J. On a result due to L.A. Caffarelli and A. Friedman concerning the asymptotic behavior of a plasma. Pitman Res. Notes in Math. 84 (1983) 214-239

[173] Sperb, R. Extension of two theorems of Payne to some nonlinear Dirichlet problems. Z. Angew. Math. Phys. 26 (1975) 721-726

[174] Sperb, R. Maximum principles and their applications. Acad. Press, New York 1981

[175] Sperner, E. Jr. Zur Symmetrisierung von Funktionen auf Sphären. Math. Z. 134 (1973) 317-327

[176] Sperner, E. Jr. Symmetrisierung für Funktionen mehrerer reeller Variablen. manuscripta math. 11 (1974) 159-170

[177] Sperner, E. Jr. Nichtlineare elliptische Differentialgleichungen in singulären Gebieten. Habilitationsschrift, Bayreuth 1979

[178] Spruck, J. Uniqueness in a diffusion model of population biology. Commun. Partial Differ. Equations 8 (1983) 1605-1620

[179] Strang, G., F. Fix An analysis of the finite element method, Prentice Hall Inc., Englewood Cliffs, N.J. 1973

[180] Strauss, W. Existence of solitary waves in higher dimensions. Commun. Math. Phys. 55 (1977) 149-162

[181] Stuart, C.A. A variational approach to bifurcation in L^p on an unbounded symmetrical domain. Math. Ann. 263 (1983) 51-59

[182] Szegö, G. On a certain kind of symmetrization and its applications. Ann. Mat. Pura Appl., IV. Ser. 40 (1955) 113-119

[183] Talenti, G. Elliptic equations and rearrangements. Ann. Sc. Norm. Sup., Pisa, IV. Ser., 3 (1976) 697-718

[184] Talenti, G. Best constant in Sobolev inequality. Ann. Mat. Pura Appl., IV. Ser. 110 (1976) 353-372

[185] Talenti, G. Nonlinear elliptic equations, rearrangements of functions and Orlicz spaces. Ann. Mat. Pura Appl., IV. Ser. 120 (1977) 159-184

[186] Talenti, G. A note on the Gauss curvature of harmonic and minimal surfaces. Pac. J. Math. 101 (1982) 477-492

[187] Talenti, G. Some estimates of solutions to Monge-Ampère type equations in dimension two. Ann. Sc. Norm. Sup., Pisa, IV. Ser. 8 (1981) 183-230

[188] Talenti, G. Linear elliptic pde's : level sets, rearrangements and a priori estimates of solutions. preprint 1983/84

[189] Talenti, G. On functions whose lines of steepest descent bend proportionally to level lines. Ann. Sc. Norm. Sup., Pisa, IV. Ser. 10 (1983) 587-605

[190] Tepper, D.F. Free boundary problem. SIAM J. Math. Anal. 5 (1974) 841-846

[191] Tepper, D.F. On a free boundary problem, the starlike case. SIAM J. Math. Anal. 6 (1975) 503-505

[192] Ting, T.W. Elastic plastic torsion problem. Arch. Ration. Mech. Anal. 25 (1967) 342-366

[193] Tolksdorff, P. Regularity for a more general class of quasilinear elliptic equations. J. Differ. Equations 50 (1983) 1627-1652

[194] Tonelli, L. Sur un problème de Lord Rayleigh. Monatsh. Math. Phys. 37 (1930) 253-280

[195] Vázquez, J.L. Symétrisation pour $u_t = \Delta\varphi(u)$ et applications. C.R. Acad. Sci., Paris. Ser. I, 295 (1982) 71-74

[196] Vázquez, J.L. Symmetrization in nonlinear parabolic equations. Port. Math. 41 (1982) 339-346

[197] Veron, L. Behaviour of solutions to nonlinear elliptic equations near a singularity of codimension 2. Pitman Res. Notes in Math. 89 (1983) 274-284

[198] Walter, W. Differential and integral inequalities. Springer-Verlag, Berlin 1970

[199] Weinberger, H.F. Symmetrization in uniformly elliptic problems, in: Studies in Math. Analysis and Related Topics, Univ. of Calif. Press, Stanford CA (1962), 424-428

[200] Weitsman, A. Spherical symmetrization in the theory of elliptic partial differential equations. Commun. Partial Differ. Equations 8 (1983) 545-561

INDEX OF EXAMPLES AND ASSUMPTIONS

SUBJECT INDEX

Vol. 1008: Algebraic Geometry. Proceedings, 1981. Edited by J. Dolgachev. V, 138 pages. 1983.

Vol. 1009: T.A. Chapman, Controlled Simple Homotopy Theory and Applications. III, 94 pages. 1983.

Vol. 1010: J.-E. Dies, Chaînes de Markov sur les permutations. IX, 226 pages. 1983.

Vol. 1011: J.M. Sigal. Scattering Theory for Many-Body Quantum Mechanical Systems. IV, 132 pages. 1983.

Vol. 1012: S. Kantorovitz, Spectral Theory of Banach Space Operators. V, 179 pages. 1983.

Vol. 1013: Complex Analysis – Fifth Romanian-Finnish Seminar. Part 1. Proceedings, 1981. Edited by C. Andreian Cazacu, N. Boboc, M. Jurchescu and I. Suciu. XX, 393 pages. 1983.

Vol. 1014: Complex Analysis – Fifth Romanian-Finnish Seminar. Part 2. Proceedings, 1981. Edited by C. Andreian Cazacu, N. Boboc, M. Jurchescu and I. Suciu. XX, 334 pages. 1983.

Vol. 1015: Equations différentielles et systèmes de Pfaff dans le champ complexe – II. Seminar. Edited by R. Gérard et J.P. Ramis. V, 411 pages. 1983.

Vol. 1016: Algebraic Geometry. Proceedings, 1982. Edited by M. Raynaud and T. Shioda. VIII, 528 pages. 1983.

Vol. 1017: Equadiff 82. Proceedings, 1982. Edited by H.W. Knobloch and K. Schmitt. XXIII, 666 pages. 1983.

Vol. 1018: Graph Theory, Łagów 1981. Proceedings, 1981. Edited by M. Borowiecki, J.W. Kennedy and M.M. Sysło. X, 289 pages. 1983.

Vol. 1019: Cabal Seminar 79–81. Proceedings, 1979–81. Edited by A.S. Kechris, D.A. Martin and Y.N. Moschovakis. V, 284 pages. 1983.

Vol. 1020: Non Commutative Harmonic Analysis and Lie Groups. Proceedings, 1982. Edited by J. Carmona and M. Vergne. V, 187 pages. 1983.

Vol. 1021: Probability Theory and Mathematical Statistics. Proceedings, 1982. Edited by K. Itô and J.V. Prokhorov. VIII, 747 pages. 1983.

Vol. 1022: G. Gentili, S. Salamon and J.-P. Vigué. Geometry Seminar "Luigi Bianchi", 1982. Edited by E. Vesentini. VI, 177 pages. 1983.

Vol. 1023: S. McAdam, Asymptotic Prime Divisors. IX, 118 pages. 1983.

Vol. 1024: Lie Group Representations I. Proceedings, 1982–1983. Edited by R. Herb, R. Lipsman and J. Rosenberg. IX, 369 pages. 1983.

Vol. 1025: D. Tanré, Homotopie Rationnelle: Modèles de Chen, Quillen, Sullivan. X, 211 pages. 1983.

Vol. 1026: W. Plesken, Group Rings of Finite Groups Over p-adic Integers. V, 151 pages. 1983.

Vol. 1027: M. Hasumi, Hardy Classes on Infinitely Connected Riemann Surfaces. XII, 280 pages. 1983.

Vol. 1028: Séminaire d'Analyse P. Lelong – P. Dolbeault – H. Skoda. Années 1981/1983. Edité par P. Lelong, P. Dolbeault et H. Skoda. VIII, 328 pages. 1983.

Vol. 1029: Séminaire d'Algèbre Paul Dubreil et Marie-Paule Malliavin. Proceedings, 1982. Edité par M.-P. Malliavin. V, 339 pages. 1983.

Vol. 1030: U. Christian, Selberg's Zeta-, L-, and Eisensteinseries. XII, 196 pages. 1983.

Vol. 1031: Dynamics and Processes. Proceedings, 1981. Edited by Ph. Blanchard and L. Streit. IX, 213 pages. 1983.

Vol. 1032: Ordinary Differential Equations and Operators. Proceedings, 1982. Edited by W.N. Everitt and R.T. Lewis. XV, 521 pages. 1983.

Vol. 1033: Measure Theory and its Applications. Proceedings, 1982. Edited by J.M. Belley, J. Dubois and P. Morales. XV, 317 pages. 1983.

Vol. 1034: J. Musielak, Orlicz Spaces and Modular Spaces. V, 222 pages. 1983.

Vol. 1035: The Mathematics and Physics of Disordered Media. Proceedings, 1983. Edited by B.D. Hughes and B.W. Ninham. VII, 432 pages. 1983.

Vol. 1036: Combinatorial Mathematics X. Proceedings, 1982. Edited by L.R.A. Casse. XI, 419 pages. 1983.

Vol. 1037: Non-linear Partial Differential Operators and Quantization Procedures. Proceedings, 1981. Edited by S.I. Andersson and H.-D. Doebner. VII, 334 pages. 1983.

Vol. 1038: F. Borceux, G. Van den Bossche, Algebra in a Localic Topos with Applications to Ring Theory. IX, 240 pages. 1983.

Vol. 1039: Analytic Functions, Błażejewko 1982. Proceedings. Edited by J. Ławrynowicz. X, 494 pages. 1983

Vol. 1040: A. Good, Local Analysis of Selberg's Trace Formula. III, 128 pages. 1983.

Vol. 1041: Lie Group Representations II. Proceedings 1982–1983. Edited by R. Herb, S. Kudla, R. Lipsman and J. Rosenberg. IX, 340 pages. 1984.

Vol. 1042: A. Gut, K. D. Schmidt, Amarts and Set Function Processes. III, 258 pages. 1983.

Vol. 1043: Linear and Complex Analysis Problem Book. Edited by V.P. Havin, S.V. Hruščëv and N.K. Nikol'skii. XVIII, 721 pages. 1984.

Vol. 1044: E. Gekeler, Discretization Methods for Stable Initial Value Problems. VIII, 201 pages. 1984.

Vol. 1045: Differential Geometry. Proceedings, 1982. Edited by A.M. Naveira. VIII, 194 pages. 1984.

Vol. 1046: Algebraic K-Theory, Number Theory, Geometry and Analysis. Proceedings, 1982. Edited by A. Bak. IX, 464 pages. 1984.

Vol. 1047: Fluid Dynamics. Seminar, 1982. Edited by H. Beirão da Veiga. VII, 193 pages. 1984.

Vol. 1048: Kinetic Theories and the Boltzmann Equation. Seminar, 1981. Edited by C. Cercignani. VII, 248 pages. 1984.

Vol. 1049: B. Iochum, Cônes autopolaires et algèbres de Jordan. VI, 247 pages. 1984.

Vol. 1050: A. Prestel, P. Roquette, Formally p-adic Fields. V, 167 pages. 1984.

Vol. 1051: Algebraic Topology, Aarhus 1982. Proceedings. Edited by I. Madsen and B. Oliver. X, 665 pages. 1984.

Vol. 1052: Number Theory. Seminar, 1982. Edited by D.V. Chudnovsky, G.V. Chudnovsky, H. Cohn and M.B. Nathanson. V, 309 pages. 1984.

Vol. 1053: P. Hilton, Nilpotente Gruppen und nilpotente Räume. V, 221 pages. 1984.

Vol. 1054: V. Thomée, Galerkin Finite Element Methods for Parabolic Problems. VII, 237 pages. 1984.

Vol. 1055: Quantum Probability and Applications to the Quantum Theory of Irreversible Processes. Proceedings, 1982. Edited by L. Accardi, A. Frigerio and V. Gorini. VI, 411 pages. 1984.

Vol. 1056: Algebraic Geometry. Bucharest 1982. Proceedings, 1982. Edited by L. Bădescu and D. Popescu. VII, 380 pages. 1984.

Vol. 1057: Bifurcation Theory and Applications. Seminar, 1983. Edited by L. Salvadori. VII, 233 pages. 1984.

Vol. 1058: B. Aulbach, Continuous and Discrete Dynamics near Manifolds of Equilibria. IX, 142 pages. 1984.

Vol. 1059: Séminaire de Probabilités XVIII, 1982/83. Proceedings. Edité par J. Azéma et M. Yor. IV, 518 pages. 1984.

Vol. 1060: Topology. Proceedings, 1982. Edited by L.D. Faddeev and A.A. Mal'cev. VI, 389 pages. 1984.

Vol. 1061: Séminaire de Théorie du Potentiel. Paris, No. 7. Proceedings. Directeurs: M. Brelot, G. Choquet et J. Deny. Rédacteurs: F. Hirsch et G. Mokobodzki. IV, 281 pages. 1984.

Vol. 1062: J. Jost, Harmonic Maps Between Surfaces. X, 133 pages. 1984.

Vol. 1063: Orienting Polymers. Proceedings, 1983. Edited by J.L. Ericksen. VII, 166 pages. 1984.

Vol. 1064: Probability Measures on Groups VII. Proceedings, 1983. Edited by H. Heyer. X, 588 pages. 1984.

Vol. 1065: A. Cuyt, Padé Approximants for Operators: Theory and Applications. IX, 138 pages. 1984.

Vol. 1066: Numerical Analysis. Proceedings, 1983. Edited by D.F. Griffiths. XI, 275 pages. 1984.

Vol. 1067: Yasuo Okuyama, Absolute Summability of Fourier Series and Orthogonal Series. VI, 118 pages. 1984.

Vol. 1068: Number Theory, Noordwijkerhout 1983. Proceedings. Edited by H. Jager. V, 296 pages. 1984.

Vol. 1069: M. Kreck, Bordism of Diffeomorphisms and Related Topics. III, 144 pages. 1984.

Vol. 1070: Interpolation Spaces and Allied Topics in Analysis. Proceedings, 1983. Edited by M. Cwikel and J. Peetre. III, 239 pages. 1984.

Vol. 1071: Padé Approximation and its Applications, Bad Honnef 1983. Prodeedings. Edited by H. Werner and H.J. Bünger. VI, 264 pages. 1984.

Vol. 1072: F. Rothe, Global Solutions of Reaction-Diffusion Systems. V, 216 pages. 1984.

Vol. 1073: Graph Theory, Singapore 1983. Proceedings. Edited by K.M. Koh and H.P. Yap. XIII, 335 pages. 1984.

Vol. 1074: E.W. Stredulinsky, Weighted Inequalities and Degenerate Elliptic Partial Differential Equations. III, 143 pages. 1984.

Vol. 1075: H. Majima, Asymptotic Analysis for Integrable Connections with Irregular Singular Points. IX, 159 pages. 1984.

Vol. 1076: Infinite-Dimensional Systems. Proceedings, 1983. Edited by F. Kappel and W. Schappacher. VII, 278 pages. 1984.

Vol. 1077: Lie Group Representations III. Proceedings, 1982–1983. Edited by R. Herb, R. Johnson, R. Lipsman, J. Rosenberg. XI, 454 pages. 1984.

Vol. 1078: A.J.E.M. Janssen, P. van der Steen, Integration Theory. V, 224 pages. 1984.

Vol. 1079: W. Ruppert. Compact Semitopological Semigroups: An Intrinsic Theory. V, 260 pages. 1984

Vol. 1080: Probability Theory on Vector Spaces III. Proceedings, 1983. Edited by D. Szynal and A. Weron. V, 373 pages. 1984.

Vol. 1081: D. Benson, Modular Representation Theory: New Trends and Methods. XI, 231 pages. 1984.

Vol. 1082: C.-G. Schmidt, Arithmetik Abelscher Varietäten mit komplexer Multiplikation. X, 96 Seiten. 1984.

Vol. 1083: D. Bump, Automorphic Forms on GL (3,IR). XI, 184 pages. 1984.

Vol. 1084: D. Kletzing, Structure and Representations of Q-Groups. VI, 290 pages. 1984.

Vol. 1085: G.K. Immink, Asymptotics of Analytic Difference Equations. V, 134 pages. 1984.

Vol. 1086: Sensitivity of Functionals with Applications to Engineering Sciences. Proceedings, 1983. Edited by V. Komkov. V, 130 pages. 1984

Vol. 1087: W. Narkiewicz, Uniform Distribution of Sequences of Integers in Residue Classes. VIII, 125 pages. 1984.

Vol. 1088: A.V. Kakosyan, L.B. Klebanov, J.A. Melamed, Characterization of Distributions by the Method of Intensively Monotone Operators. X, 175 pages. 1984.

Vol. 1089: Measure Theory, Oberwolfach 1983. Proceedings. Edited by D. Kölzow and D. Maharam-Stone. XIII, 327 pages. 1984.

Vol. 1090: Differential Geometry of Submanifolds. Pro 1984. Edited by K. Kenmotsu. VI, 132 pages. 1984.

Vol. 1091: Multifunctions and Integrands. Proceedings, 19 by G. Salinetti. V, 234 pages. 1984.

Vol. 1092: Complete Intersections. Seminar, 1983. Edited b and R. Strano. VII, 299 pages. 1984.

Vol. 1093: A. Prestel, Lectures on Formally Real Fields. XI, 1 1984.

Vol. 1094: Analyse Complexe. Proceedings, 1983. Edité pa R. Gay et Nguyen Thanh Van. IX, 184 pages. 1984.

Vol. 1095: Stochastic Analysis and Applications. Proceedin Edited by A. Truman and D. Williams. V, 199 pages. 1984.

Vol. 1096: Théorie du Potentiel. Proceedings, 1983. Edit Mokobodzki et D. Pinchon. IX, 601 pages. 1984.

Vol. 1097: R.M. Dudley, H. Kunita, F. Ledrappier, École Probabilités de Saint-Flour XII – 1982. Edité par P.L. Henn 396 pages. 1984.

Vol. 1098: Groups – Korea 1983. Proceedings. Edited by A and B.H. Neumann. VII, 183 pages. 1984.

Vol. 1099: C.M. Ringel, Tame Algebras and Integral Quadrati XIII, 376 pages. 1984.

Vol. 1100: V. Ivrii, Precise Spectral Asymptotics for Elliptic O Acting in Fiberings over Manifolds with Boundary. V, 237 1984.

Vol. 1101: V. Cossart, J. Giraud, U. Orbanz, Resolution of Singularities. VII, 132 pages. 1984.

Vol. 1102: A. Verona, Stratified Mappings – Structure and T lability. IX, 160 pages. 1984.

Vol. 1103: Models and Sets. Proceedings, Logic Colloquium Part I. Edited by G.H. Müller and M.M. Richter. VIII, 484 pages

Vol. 1104: Computation and Proof Theory. Proceedings, Colloquium, 1983, Part II. Edited by M.M. Richter, E. Börger, W schelp, B. Schinzel and W. Thomas. VIII, 475 pages. 1984.

Vol. 1105: Rational Approximation and Interpolation. Procee 1983. Edited by P.R. Graves-Morris, E.B. Saff and R.S. XII, 528 pages. 1984.

Vol. 1106: C.T. Chong, Techniques of Admissible Recursion T IX, 214 pages. 1984.

Vol. 1107: Nonlinear Analysis and Optimization. Proceedings, Edited by C. Vinti. V, 224 pages. 1984.

Vol. 1108: Global Analysis – Studies and Applications I. Edi Yu.G. Borisovich and Yu.E. Gliklikh. V, 301 pages. 1984.

Vol. 1109: Stochastic Aspects of Classical and Quantum Sy Proceedings, 1983. Edited by S. Albeverio, P. Combe and M. Si Collin. IX, 227 pages. 1985.

Vol. 1110: R. Jajte, Strong Limit Theorems in Non-Commu Probability. VI, 152 pages. 1985.

Vol. 1111: Arbeitstagung Bonn 1984. Proceedings. Edited by F. bruch, J. Schwermer and S. Suter. V, 481 pages. 1985.

Vol. 1112: Products of Conjugacy Classes in Groups. Edit Z. Arad and M. Herzog. V, 244 pages. 1985.

Vol. 1113: P. Antosik, C. Swartz, Matrix Methods in Analysi 114 pages. 1985.

Vol. 1114: Zahlentheoretische Analysis. Seminar. Herausgeg von E. Hlawka. V, 157 Seiten. 1985.

Vol. 1115: J. Moulin Ollagnier, Ergodic Theory and Statistical Me ics. VI, 147 pages. 1985.

Vol. 1116: S. Stolz, Hochzusammenhängende Mannigfaltigk und ihre Ränder. XXIII, 134 Seiten. 1985.